Studies in Economic History

Aims and Scope

This series from Springer provides a platform for works in economic history that truly integrate economics and history. Books on a wide range of related topics are welcomed and encouraged, including those in macro-economic history, financial history, labor history, industrial history, agricultural history, the history of institutions and organizations, spatial economic history, law and economic history, political economic history, historical demography, and environmental history.

Economic history studies have greatly developed over the past several decades through application of economics and econometrics. Particularly in recent years, a variety of new economic theories and sophisticated econometric techniques—including game theory, spatial economics, and generalized method of moment (GMM)—have been introduced for the great benefit of economic historians and the research community.

At the same time, a good economic history study should contribute more than just an application of economics and econometrics to past data. It raises novel research questions, proposes a new view of history, and/or provides rich documentation. This series is intended to integrate data analysis, close examination of archival works, and application of theoretical frameworks to offer new insights and even provide opportunities to rethink theories.

The purview of this new Springer series is truly global, encompassing all nations and areas of the world as well as all eras from ancient times to the present. The editorial board, who are internationally renowned leaders among economic historians, carefully evaluate and judge each manuscript, referring to reports from expert reviewers. The series publishes contributions by university professors and others well established in the academic community, as well as work deemed to be of equivalent merit.

More information about this series at http://www.springer.com/series/13279

Minoru Sawai
Editor

The Development of Railway Technology in East Asia in Comparative Perspective

Springer

Editor
Minoru Sawai
Nanzan University
Nagoya
Japan

ISSN 2364-1797 ISSN 2364-1800 (electronic)
Studies in Economic History
ISBN 978-981-10-4903-3 ISBN 978-981-10-4904-0 (eBook)
DOI 10.1007/978-981-10-4904-0

Library of Congress Control Number: 2017943093

Printed on acid-free paper

This Springer imprint is published by Springer Nature
The registered company is Springer Nature Singapore Pte Ltd.
The registered company address is: 152 Beach Road, #21-01/04 Gateway East, Singapore 189721, Singapore

Preface

This is the first book to examine the process of railway development in Japan, Korea, Taiwan, and China from historical and comparative perspectives. Moreover, it discusses and compares the East Asian experiences of railway development with cases in Germany, which had been a mainstay of railway development in Europe. Railway technology is a set of technology on locomotives, rolling stock, construction of railways, bridges, tunnels, electricity, metallurgy, and so on. Discussion on railway technology is, therefore, examination of industrial technology as a whole in each national economy.

After the opening of Japan in the mid-nineteenth century, the country achieved import substitution of locomotives in half a century. This book explores the social capability of Meiji Japan to overtake the advanced countries in railway technology. Parallel with the expansion of the Japanese empire, a large team of engineers constructed and operated the colonial government railways of Taiwan and Korea and South Manchuria Railway. This book clearly outlines the education and training of these engineers. The management capabilities of the colonial railways and South Manchuria Railway were transferred to the postwar period. The development of railway in Taiwan, for example, was an integration of "legacy" of colonial times, an influence of Chinese officers or engineers who transferred from Mainland China with Chiang Kai-shek Government, and profound impact of the US aid. Through "dieselization" in postwar years, locomotives were changed from the Japanese-made to American-made in the purchase process under the US aid. These dramatic East Asian experiences of railway technology development are compared with European cases, mainly German railways. In comparison with the experience of the Japanese nationalization of railways, the target of comparison should not be the "Nationalisierung" of railways in German empire but the "State-Owing" of railways in "Land" (local state) such as Kingdom of Prussia.

Five authors of each chapter in this book took part in the session titled "the development of railway technology in East Asia" as presenters and a commentator at the annual meeting of Business History Society of Japan in October 2015. After the examination of many comments given at the session, and brushing up the contents of each presentation, we again made presentation at the session of the First World Congress on Business History held in Bergen, Norway, in August 2016. Through the discussion at the two sessions in Japan and Norway, we could recognize the significance of comparative history on railway technology. This book is the first step for us to proceed with comparative history of railways that exemplified the industrial level of each national economy.

Finally, I must thank every commentators and participants at the sessions who delivered valuable comments and discussion to help elaborate the contents of this book.

Nagoya, Japan Minoru Sawai

Contents

Contributors

Minoru Sawai is a Professor at Nanzan University and President of Business History Society of Japan and of the Forum for Entrepreneurial Studies. He obtained a Ph.D. in economics from Osaka University in 1998. His main research field is the economic and business history of modern Japan. His publications include *Nihon Tetsudō Sharyō Kōgyōshi* (History of the Japanese Locomotives and Rolling Stock Industry)(Nihon Keizai Hyōronsha, 1998) and *Maza Mashin no Yume: Nihon Kōsaku Kikai Kōgyōshi* (Dream of Mother Machines: History of the Japanese Machine Tool Industry)(The University of Nagoya Press, 2013).

Naofumi Nakamura is a Professor at Institute of Social Science, the University of Tokyo. He obtained a Ph.D. in history from Kyushu University in 1997. His main research field is the economic and business history of modern Japan. His publications include *Chihō karano Sangyō-kakumei* (Reconsidering the Japan's industrial revolution: A local perspectives) (The University of Nagoya Press, 2010) and *Umi wo wataru Kikansha* (Trading locomotives: The first globalization and the development of Japan's railways, 1869–1914) (Yoshikawa Kōbun-kan, 2016).

Chaisung Lim is a Professor in the College of Economics, Rikkyo University. He obtained a Ph.D. in economics from the University of Tokyo in 2002. His main research field is the economic history of East Asia. His publications include *Senjikeizai to Tetsudō* (A Wartime Economy and Railroad Management)(University of Tokyo Press, 2005) and *Kahokukōtsū no Nitchūsensō-shi* (The Sino-Japanese War History of North China Railway Company)(Nihon Keizai Hyōronsha, 2016).

Lung-Pao Tsai is a Professor of Department of History, National Taipei University. He obtained a Ph.D. in history from National Taiwan Normal University in 2007. His main research field is the economic and business history of modern Taiwan. His publications include *Tuīdòng shídài de jùlún: Rìzhì zhōngqí de táiwān guóyǒu tiělù (1910–1936)* (The Development of Taiwan Railway in the

Japanese-ruled Period, 1910–1936) (Táiwān shūfáng, 2004) and *Zhímín tǒngzhì zhī jīchǔ gōngchéng-rì zhì shíqí táiwān dàolù shìyè zhī yánjiū 1895–1945* (The Infrastructure of the Colonial Rule: A Case Study of Taiwan's Road Business in the Japanese-ruled Period, 1895–1945) (Guólì táiwān shīfàn dàxué lìshǐ xué xì, 2008)

Ayumu Banzawa is a Professor of the Graduate School of Economics, Osaka University. He obtained a Ph.D. in economics from Osaka University in 2001. His main research field is the economic and business history of modern Germany. His publications include Doitsu-kogyô-ka ni okeru Tetsudôgyô (Railways in German Industrialization) (Yûhikaku, 2006) and *The Development of Western Business History Research in Japan: With reference to the Simultaneity with Research in Western Countries* (with Makiko Hino and Hiroko Sadatô) in *Japanese Research in Business History*, 32.

Chapter 1
Introduction

Minoru Sawai

Abstract It was a bureau of manufacturing of the Japanese National Railway (JNR) that decisively controlled the development of locomotives and rolling stock in modern Japan. Largely depending on higher technical educational organizations outside Japan proper in the supply of engineers, railways in the Japanese empire were organized into the structure of ellipse with two centers of the JNR and South Manchuria Railway (SMR). The process of railway technology development in Japan before the First World War could be divided into four stages: (1) transfer of railway systems from the UK by the Japanese government in the 1870s and the 1880s; (2) acquisition of railway materials from the US and Germany and diversification of technology in the 1890s; (3) standardization and convergence of technology through the merger and nationalization of railway networks in the 1900s; and (4) the development of original technology by the government railway in the 1910s. SMR emphasized a technology transfer from Western world based on management stabilities and tried to manufacture its own railroad cars by itself, then participated in the world competition of the limited express passenger train in the 1930s. On the other hand, postwar Taiwan Railway Administration recovered from the damage of the war and developed from the old power to "dieselization" during the postwar to the US aid period, and locomotives were changed from the Japanese-made to American-made in the purchase process under the US aid. Lastly from the perspectives of comparative history, we compare the nationalization process of railways in Japan and Germany. As characteristics of expansion process of the "national" railways in Kingdom of Prussia, we could confirm two points: tradition of federated decentralization of authority and halfway centralization of power that was shown as absence of functional equivalent to Tetsudô In (Agency of Railways) in Japan.

Keywords Railway technology · Steam locomotives · East Asia · Japan National Railways · South Manchuria Railway · Taiwan railway administration · Germany

M. Sawai (✉)
Faculty of Business Administration, Nanzan University, 18 Yamazato-cho, Showa-ku, Nagoya 466-8673, Japan
e-mail: msawai@nanzan-u.ac.jp

M. Sawai (ed.), *The Development of Railway Technology in East Asia in Comparative Perspective*, Studies in Economic History,
DOI 10.1007/978-981-10-4904-0_1

Construction and maintenance of railways is supported by many technical fields such as locomotives, rolling stock, tracks, bridges, stations, electricity, cement, lumbers, metal materials, and machine tools. We would like to figure out the leads in order to understand the characteristics of railway technology that supported the expansion of the Japanese empire, and the development of railways in postwar East Asian countries that partly inherited the technological "legacy" from colonial period.

So many academic works on railway technology have been accumulated. Confining to the history of locomotives, for example, we could understand dynamic evolution and elaboration process in American experiences. John K. Brown demonstrated the competitive edge of Baldwin Locomotive Works before the First World War (Brown 1995). Albert Churella clearly showed the transition process from steam locomotives to diesel locomotives in the US (Churella 1998).

Minoru Sawai surveyed the development of the Japanese steam locomotives and rolling stock industry from Meiji era to the end of the Pacific War (Sawai 1998). Naofumi Nakamura analyzed the competition of sales of locomotives by UK, USA, and Germany in the world market, in particular in Japan, using many previously untapped resources (Nakamura 2016).

In Chap. 2 we examine the supply structure of engineers related to the locomotive and rolling stock in the Japanese empire. It was a bureau of manufacturing of the Japanese National Railway (JNR) that decisively controlled the development of locomotives and rolling stock. We could see the tight-knitted relations between the South Manchuria Railway Company (SMR), the Lushun Institute of Technology (LIT) and the South Manchuria Technical College (SMTC) in the supply of engineers. Largely depending on higher technical educational organizations outside Japan proper in the supply of engineers, railways in the Japanese empire were organized into the structure of ellipse with two centers of the JNR and South Manchuria Railway Company.

Among the technological activities led by JNR, the Conference on Rolling Stock (*Sharyô Kenkyûkai*) was significant. The group of JNR engineers from the rolling stock division of the manufacturing bureau formed core of *Sharyô Kenkyûkai*, augmented by engineers from SMR, government railways in Taiwan and Korea, private railway companies, materials and parts producers, and designated factories. The conference, where collaborative research could be evaluated via exchanges between users and builders of rolling stock and locomotives, contributed to the technological upgrading of the Japanese Empire's rolling stock industry. The collaborative research between users and designated manufacturers able to produce high-quality products was one aspect of the efficient R&D adopted by a latecomer to industrialization—i.e., Imperial Japan. In this method of selecting qualified manufacturers, the state—the largest buyer—determined which manufacturers were candidates for optional contracts. For non-member firms excluded from collaborative research, the closed system of R&D composed of JNR and SMR was a kind of "crony innovation system."

The main purpose of Chap. 3 is to clarify the process of railway technology development in Japan during the Meiji era, focusing on the role of railway engineers and their technical imitation. This chapter explores the characteristics of railway technology development in East Asia through the example of Japan, which was the first East Asian country to develop original railway technology.

The process of railway technology development in Japan could be divided into four stages: (1) transfer of railway systems from the UK by the Japanese government in the 1870s and the 1880s; (2) acquisition of railway materials from the US and Germany and diversification of technology in the 1890s; (3) standardization and convergence of technology through the merger and nationalization of railway networks in the 1900s; and (4) the development of original technology by the government railway in the 1910s.

Around the turn of the century, efforts to organize institutes of higher education and vocational schools to cultivate engineers were advanced, and a full-fledged effort was begun to expand the ranks of Japanese engineers. Factories affiliated with railway companies and independent rolling stock builders began to duplicate the foreign-made locomotives. It became necessary to regulate the capacity and rail weight of individual railways as well as the strength of couplers after the Sino-Japanese War. The advancement of standardization experienced a sudden boost with the nationalization of railways in 1906–1907, which led to the convergence of railway technology.

In 1914, it succeeded in developing a locomotive conforming to Japanese standards that was modeled after a German-made locomotive but incorporated original design elements. Thereafter, various locomotives conforming to Japanese standards based on original designs were successively developed, and Japan entered the age of locomotive building based on original technology.

Chapter 4 examines the railway business of the Mainland China established by the Japanese initiative after Russo-Japanese War in 1904 and 1905, focusing mainly on track facilities and railroad cars, to figure out how the technology transfer, its indigenization and the further technological process were realized.

According to the result of the Russo-Japanese War, the southern line of the Chinese Eastern Railway of Russian Empire has begun to function as a railway network of the Japanese Empire with Gyeongbu and Gyeongui Lines of Korean Peninsula. The network of SMR was set up around the northeastern part of China with the railroad construction, and it goes without saying that it greatly contributed to the maintenance of the imperial order. However, SMR had no option but to buy the railroad cars from abroad, because the foreign railroad management of a normal-gauge railway was the first experience for Japanese and rolling stock and machine industry in Japan did not have enough technological capacities to support it.

Nevertheless, when the technology transfer was implemented, SMR began to pursuit its own technology. Succeeding early in relatively much passenger and freight traffic and realized high productivity, SMR was able to have much sounder profit structure with Fushun coal mine than JNR which was stable in its management just after the establishment as well as Korean governmental railways operated by the Government-General of Korea which was troubled by management instability until the 1930s when the colonial industrialization evolved on a large scale.

SMR emphasized a technology transfer from Western world based on management stabilities and tried to produce its own railroad cars by itself, then participated in the world competition of the limited express passenger train in the 1930s. This involved adjacent Korean governmental railways. The service of the limited express passenger train reached North Manchurian via Korea from the Japan inland with the establishment of Manchukuo in 1932 and spread to Beijing after the outbreak of Second Sino-Japanese War in 1937.

Chapter 5 examines how the postwar Taiwan Railway Administration (TRA) normalization and further development were achieved by the efforts of Chiang Kai-shek Government and Taiwanese employees, and the US aid which started from 1950. During this process, it was noteworthy how Taiwan's railway system would be integrated. Would it be integrated towards the Chinese style because of the technical officers of Chiang Kai-shek Government, or American style due to the US aid, or maintain Japanese style as before the war?

The meaningful problem involved not only the economic issue but also the postwar Taiwan's politics and diplomacy to a large extent. With the above-mentioned problem in the mind, we try to investigate the most technological problem of "power renovation" in railway development by delving into how the TRA recovered from the damage of the war and developed from the old power to "dieselization" during the postwar to the US aid period, and why the locomotives were changed from the Japanese-made to American-made in the purchase process under the US aid.

Chapter 5 analyzes the effects and influence of the "dieselization policy". "Renovation" requires funds and technology. At that time, the foreign funds made the most part of the capital composition, and the "foreign funds" often accompanied "technology". Period from 1950 to 1965 was the US-aid era, and the technologies accompanying the US aid included American consulting companies, railway professional advisers, advanced machinery purchases, and so on. How they did affect the TRA profoundly is an important research question in economic history of East Asia.

Finally the research question of Chap. 6 is why German empire (Reich) before the First World War could not accomplish the nationalization of railways within the empire. In comparison with the experience of the Japanese nationalization of railways, Chap. 6 insists the target of comparison is not the "Nationalisierung" of railways in German empire but the "State-Owing" of railways in "Land" (local state) such as Kingdom of Prussia. The targets of research are analyses on variety on "Partikularismus" of local states which seems to hamper the nationalization of railways in German empire, and on difficulties of centralization of executive agencies as preconditions of unification of railways.

It is very difficult to execute comparative research on unification of local railways managed and operated by independent sovereign (or semi-sovereign) states in German empire and state-owing through purchases of private railways by a sovereign state in the case of Japan. As characteristics of expansion process of the "national" railways in Kingdom of Prussia, we could confirm two points: tradition of federated decentralization of authority and halfway centralization of power that was shown as absence of functional equivalent to Tetsudô In (Agency of Railways) in Japan. In this sense presence or absence of federated system had a decisive impact on "nationalization" of railways.

Furthermore it is interesting that operative organizations of Japan and state-owning railways in Kingdom of Prussia had arrived at close points through different paths; central bureau of Prussian railways was newly added to organizational structure based on regional decentralization, while technological limitation of centralized operation of railways gave rise to organizations of regionally decentralized system to a certain degree in Japan.

References

Brown, John K. 1995. *The baldwin locomotives works: 1831–1915*. Baltimore: John Hopkins University Press.

Churella, Albert. 1998. *From steam to diesel: management customs and organizational capability in the twentieth-century American locomotive industry*. Princeton: Princeton University Press.

Sawai, Minoru. 1998. *Nihon tetsudô sharyô kôgyô shi (History of Japanese locomotives and rolling stock industry)*. Tokyo: Nihon Keizai Hyôron Sha.

Nakamura, Naofumi. 2016. *Umi wo wataru kikansha: kindai Nihon no tetsudô hatten to guro-baruka (Trading locomotives: The first globalization and the development of Japan's Railways, 1890–1914)*. Tokyo: Yoshikawa Kôbunkan.

Chapter 2
Railway Engineers of the Japanese Empire and the Significance of Collaborative R&D Activities

Minoru Sawai

Abstract Lushun Institute of Technology (LIT) and South Manchuria Technical College (SMTC) played a great role in supplying engineers for colonial government railways in Taiwan and Korea, South Manchuria Railway (SMR) , and rolling stock and locomotive companies in Dalian and Seoul. Dependent on educational organizations, such as LIT and SMTC with regard to the supply of engineers, railways within the Japanese empire were organized into an ellipse with JNR and SMR at its apex. The collaborative research activities demonstrated in the case of Conference on Rolling Stock between users and designated manufacturers able to produce high-quality products was one aspect of the "efficient" R&D adopted by a latecomer to industrialization—i.e., Imperial Japan. However, in this method of selecting qualified manufacturers, the state—the largest buyer—determined which manufacturers were candidates for optional contracts. For non-member firms excluded from collaborative research, this was the closed system of R&D activities.

Keywords JNR · SMR · Conference on rolling stock · Technical committee on rolling stock

2.1 Introduction

During the Meiji period (1868–1912), Japan imported steam locomotives and railway technologies from the UK, USA, and Germany. During Japan's road to nationalization in 1906 and 1907, the country's railway engineers had to unify those disparate imported technologies, create their own technologies, or import substitutes. To do so, Japanese National Railways (JNR) adopted a "designated factory system" in 1912 to secure privately built rolling stock and locomotives from three factories: Kisha Seizō, Nippon Sharyō Seizō, and Kawasaki Shipyard (Amano

M. Sawai (✉)
Faculty of Business Administration, Nanzan University, 18 Yamazato-cho, Showa-Ku, Nagoya 466-8673, Japan
e-mail: msawai@nanzan-u.ac.jp

M. Sawai (ed.), *The Development of Railway Technology in East Asia in Comparative Perspective*, Studies in Economic History,
DOI 10.1007/978-981-10-4904-0_2

Factory was added later). The system prohibited purchase of imported locomotives and ended the manufacture of new rolling stock at JNR's factories.

The designated factory system established a demarcation between Japan's private and public sectors and shaped the trajectory of Japan's rolling stock industry afterwards: JNR's factories would concentrate on repair, and private companies would manufacture rolling stock and locomotives. The designated factory system for rolling stock and locomotives was one of the several similar systems established by public sector entities, including the military and Ministry of Telecommunication. This system nurtured a select few private manufacturers who were instrumental in the industrialization of modern Japan. It was a key feature in Japan's experience and the industrial development of developing economies in the 20th century.

JNR influenced the development of railway technology in Japan and elsewhere. Colonial governments in Taiwan, Korea, and south Sakhalin built railways generally operated by bureau of transportation of Taiwan Sōtokufu (operated under the Governor General of Taiwan), railway bureau of Chōsen Sōtokufu (operated under the Governor General of Korea) and Sakhalin Agency. Colonial government railway factories were established in Taiwan and Korea for both repair and manufacturing of new rolling stock and locomotives. In 1906, the South Manchuria Railway (SMR) was established, and the Shahekou railway factory was built in Dalian. Colonial government railways and the SMR had remained close to their designated factories over a long period, as had JNR, which mentored colonial governments and the SMR in maintaining relationships with designated factories. Designated factories were not limited to private rolling stock enterprises in Japan. *Ryūzan Kōsaku* in Korea and *Dairen Kikai Seisakusho* in Dalian were important private designated factories in their regions.

After the Japanese Empire collapsed in 1945, Japanese employees of colonial government railways and the SMR began to return to Japan. Most could return relatively quickly, except for a small number of engineers. Railway workers and engineers on site operated and maintained former colonial government railways and the SMR. For example, not only the "legacy" of colonial times but also Nationalist Party-related engineers who came from the continent and US foreign aid influenced the development of railway technology in postwar Taiwan. Postwar railways in Taiwan groped their way toward "Taiwanese" railway technology under these circumstances.

To consider the development of railway technologies in East Asia in the 20th century, this chapter examines the historical significance of technological alternatives under imperialism; how workers, engineers, and managers were supplied; how collaborative R&D activities organized by the JNR worked in the development of railway technology.

This chapter first examines the allocation of railway engineers, especially those involved in manufacturing rolling stock and locomotives, who advanced railway development in the Japanese Empire. Latter portions of this chapter detail the significance of collaborative R&D supported by engineers within the Japanese Empire, particularly the Conference on Rolling Stock and Technical Committee on Rolling Stock.

2.2 Engineers of the Ministry of Railways, South Manchuria Railway, and Colonial Government Railways

2.2.1 Bureau of Manufacturing Within the Ministry of Railways

The manufacturing bureau within the Ministry of Railways technologically controlled Japan's rolling stock technology.[1] Its first chief—Yasujirō Shima (from June 1915 to April 1918 in office)—was followed by seven successors: Seiji Takasu, Shōhachi Akiyama, Okiie Yamashita, Kiichi Asakura, Hisaji Kii, Shinsaku Tokunaga, and Kingo Mukasa. The bureau was reorganized under the Ministry of Transportation and Telecommunication in 1943 (Sawai 2015, 45). All eight chiefs were graduates of the Department of Engineering at Tokyo Imperial University. Initially, the manufacturing bureau comprised three divisions: factory, rolling stock, and electricity. The electricity division acquired bureau status in 1920, and the division of machinery was added in 1927. Within the manufacturing bureau, the division of rolling stock planned, designed, and executed "manufacturing and remodeling of rolling stock."(Tetsudō Daijinkanbō Jinjika 1934, 4).

After retiring from the Ministry of Railways in 1919, Shima was named the director of the SMR. In 1934, Shima became the president of Kisha Seizō, the most famous designated factory. Takasu was appointed the president of Hatsudōki Seizō and an advisor to Nippon Air Brake (Nihon Kōtsū Kyōkai 1972, 216). Akiyama, who retired in 1929, was appointed managing director and vice president at Nippon Sharyō Seizō, becoming the president of Manshū Sharyō in Manchuria. Yamashita retired in 1933 and was named the director of Nissan Motor and Hitachi Seisakusho. After retiring in 1936, Asakura was appointed the managing director of Kisha Seizō and Kii was appointed the managing director of Shibaura Kyōdō Kōgyō. Tokunaga was appointed the managing director of Nippon Sharyō Seizō. Mukasa was appointed the president of Kawasaki Heavy Industries (Kawasaki Shipyard changed its name in 1939) (Sawai 2015, 46–50).

The custom of instating chiefs of JNR's manufacturing bureau as directors at designated factories or related companies is called *amakudari* (descent from heaven). The custom deepened relationships between JNR and designated factories. *Amakudari* was also the custom among other ministries of Japan's army and navy. Former chiefs of JNR's manufacturing bureau were often appointed as board members of academic associations, such as the Japan Society of Mechanical Engineers. They were key in promoting cooperation among industry, government, and universities.

[1]Names of government offices responsible for national railways changed frequently until the mid Meiji period. In 1908, just after nationalization, the Tesudō-in (Agency of Railways) was established, followed by Tetsudō-shō (Ministry of Railways) in 1920. This chapter consistently uses the term "JNR" to indicate the continuity of these government offices.

Table 2.1 presents the members of the rolling stock division of JNR's manufacturing bureau. We could confirm 11 engineers and 15 assistant engineers. Ten of the engineers were engineering graduates of Tokyo Imperial University, and one was a graduate from the Department of Engineering at Kyushu Imperial University. The rolling stock division of JNR's manufacturing bureau and the naval ship headquarters absorbed large numbers of elite engineering graduates from imperial universities, particularly Tokyo Imperial University.[2]

Hideo Shima, first-born son of Yasujirō Shima and a 1925 engineering graduate from Tokyo Imperial University, was appointed assistant engineer of the rolling stock division of the manufacturing bureau in 1927 after a training course at JNR's Ômiya factory(Takahashi 2000, 18.). He was promoted to engineer status in 1928, became chief designer of C10 steam locomotives (SL) in 1930 at age 29, and remained in that position to design types C54 to D51, one of the most excellent engines in the history of Japan's SLs (Takahashi 2000, 19–20). Hideo Shima went abroad as a researcher from February 1936 to December 1937, then was appointed manager of locomotives at the Takatori factory within the Osaka railway bureau in January 1938. He became the superintendent of Hamamatsu factory in August 1941 and director and chief of JNR's manufacturing bureau in June 1949 (Tetsudōshi Gakkai 2013, 227–228).[3]

Senichirō Hosokawa (Table 2.1) was appointed assistant engineer of the rolling stock division of the manufacturing bureau in 1932. He was promoted to chief engineer for C56, C57, and C58 type SLs during the war, and type D52, a large locomotive that pulled freight cars. According to Hosokawa, teams needed more than 20 engineers, assistant engineers, and staff to design boilers, cylinders, wheels, and drives. Designing a new SL took about one year (Takahashi 2000, 20). An engineer concerned should be a project manager of design of new steam locomotives.

Alongside engineers and assistant engineers, the manufacturing bureau employed many technical assistants (*Gijutsu Yatoi* or *Koin*) (Table 2.2). Most were promoted to assistant engineers in charge of developing SLs. For example, Kenjirō Katayama (Table 2.1) was chief engineer for machine parts in construction of SL types EF10, EF11, EF12, and EF14, early stage locomotives that pulled freight cars. Mitsuyuki Fukasawa served that role for bogies and brakes, as did Toyokazu Maki for electrical parts (Table 2.2) (Uninscribed 1965, 16).

Before being transferred to the manufacturing bureau in Tokyo, Toyokazu Maki had worked at a local operation site (Uninscribed 1957, 66). Engineering graduates from universities and technical colleges who also had been trained in workshops advanced R&D and manufacturing processes.

---

[2]The Japanese navy brought in "entrusted students" from imperial universities every year and gave them scholarships. After graduating, they were required to undergo a year of military training and were appointed as lieutenants, called "Bukan" or "Gijutsu shikan" (naval technical officers). Graduates of universities and technical colleges who did not receive military training continued their careers as *Bunkan* or naval engineers.

[3]He returned to JNR as a chief engineer in 1955 and in March 1957 announced plans for the bullet train (*Shinkansen*) to link Tokyo and Osaka in three hours (Tetsudōshi Gakkai 2013, 228).

Table 2.1 Engineers of rolling stock division of manufacturing bureau and research institute of JNR (15th August, 1934)

Division	Name	Affiliated colleges and universities department	Major	Year of graduation
Chief of rolling stock division of manufacturing bureau	Shinsaku Tokunaga	Tokyo Imperial University, Engineering	Machinery	1915
Engineer	Shinsuke Hashimoto	Tokyo Imperial University, Engineering	Machinery	1921
〃	Zentarō Yamamoto	Kyushu Imperial University, Engineering	Electricity	1919
〃	Takeshi Gotō	Tokyo Imprerial University, Engineering	Machinery	1922
〃	Yasuto Kōzu	〃	〃	1921
〃	Sukeshige Taga	〃	〃	1923
〃	Hideo Shima	〃	〃	1925
〃	Hideo Yoshizawa	〃	Metallurgy	1925
〃	Tatsuji Satake	〃	Machinery	1926
〃	Taro Tanaka	〃	〃	1928
〃	Tomoshige Sekiguchi	〃	〃	1928
〃	Akimasa Kitabatake	〃	〃	1929
Assistant Engineer	Hironosuke Sekine	Kyoto Imperial University	〃	1918
〃	Atsutoshi Yokota	Waseda University, Science and Engineering	〃	1923
〃	Tōru Yokoyama	〃	〃	1923
〃	Tarô Yasaka	〃	Electricity	1924
〃	Keizō Hiyama	Meiji Technical College	Machinery	1923
〃	Kichiji Yoshida	Nagoya Technical College	〃	1922
〃	Kenjiro Katayama	Tokyo Imperial University, Engineering	〃	1929
〃	Mamoru Shibata	Nihon Daigaku Technical College	〃	1923
〃	Sakae Fujiwara	Yokohama Technical College	〃	1924
〃	Yasunosuke Sugano			
〃	Takeshi Ōno			
〃	Kyūzo Oka	Kobe Technical College	Electricity	1925
〃	Shōichi Yokobori	Tokyo Imperial University, Engineering	Machinery	1930

(continued)

Table 2.1 (continued)

Division	Name	Affiliated colleges and universities department	Major	Year of graduation
〃	Hideo Mori	Tokyo Technical College	〃	1926
〃	Senichirō Hosokawa	Tokyo Imperial University, Engineering	〃	1931
Director of research institute	Ryūji Yamada	Tokyo Imperial University, Engineering	Civil engineering	1911
Chief of first section	Tadakazu Chō	Tokyo Imperial University, Engineering	Applied chemistry	1908
Engineer of first section	Kinpei Yoshida	〃	〃	1914
〃	Tōichirô Amano	Tokyo Imperial University, Agriculture	Forestry	1916
〃	Kanehiro Sugayama			
〃	Fusatoshi Nozawa	Tohoku Imperial University, Science	Chemistry	1917
〃	Seiichi Hirose	Tokyo Imperial University, Engineering	Applied chemistry	1919
〃	Hidetoshi Matsunami	Kyoto Imperial University, Engineering	Metallurgy	1921
〃	Tadao Nakamura	Tokyo Imperial University, Agriculture	Forestry	1924
〃	Takashi Tamura	Tohoku Imperial University, Engineering	Chemical engineering	1914
Chief of second section	Ryūji Yamada			
Engineer of second section	Keiji Fujita	Tokyo Imperial University, Engineering	Machinery	1911
〃	Masuhiro Suzuki	Tohoku Imperial University, Science	Physics	1921
〃	Shōji Ikeda	Tohoku Imperial University, Engineering	Machinery	1923
〃	Kuraji Musashi	〃	〃	1923
〃	Haruhiko Shibata	Tokyo Imperial University, Engineering	Weapon manufacture	1924
Assistant engineer of second section	Yoshitarō Kimura	Waseda University, Science and Engineering	Machinery	1917
〃	Matsusaburō Wakasugi			
〃	Motomu Ishida	Tokyo Imperial University, Engineering	Metallurgy	1928
〃	Masaki Kubo	Waseda University, Science and Engineering	Machinery	1924
〃	Kō Chiba			

(continued)

Table 2.1 (continued)

Division	Name	Affiliated colleges and universities department	Major	Year of graduation
〃	Tōzō Ōiwa	Kanazawa Techcal College	Machinery	1928
〃	Masao Koibuchi	Tokyo Technical College	Electrical chemistry	1928
〃	Hikogorō Kobayashi			
〃	Yoshijirō Kunieda	Nihondaigaku Technical College	Machinery	1932
Chief of third section	Hisao Nakayama	Tokyo Imperial University, Engineering	Electricity	1913
Engineer of third section	Sadakazu Hattori	Tohoku Imperial University, Science	Physics	1923
〃	Shizuo Nakamura	Kyushu Imrerial University, Engineering	Electricity	1927
Chief of fourth section	Masanori Numata	Tokyo Imperial University, Engineering	Civil engineering	1919
Engineer of fourth section	Juichirō Nakahara	Kyushu Imrerial University, Engineering	〃	1921
〃	Kan Watanabe	Tokyo Imperial University, Science	Geology	1923
〃	Gonbe Inaba	Tokyo Imperial University, Engineering	Civil engineering	1923
〃	Minoru Uchiyama	〃	〃	1927
Chief of fifth section	Fumisaburō Sawa	Tokyo Imperial University, Science	Physics	1907

Source Tetsudō Daijinkanbō Jinjika (1934), Gakushikai (each year), and Nikkan Kōgyō Shinbunsha (1934)

Notes (1) Engineers from Shinsaku Tokunaga to Senichirô Hosokawa were in the rolling stock division of manufacturing bureau, and engineers from Ryûji Yamada to Fumisaburô Sawa were in the research institute of JNR. Only the second section of the institute shows assistant engineers

(2) Assistant engineers of rolling stock division of manufacturing bureau are shown only those of the second and the third grades

(3) The blank is unknown

JNR trained workers at in-house schools (*Tetsudō Kyōshūjo*) that prepared them to become assistant engineers and engineers. After 1922, the curriculum included railway middle school (five years of night school), ordinary in-house training at each local railway bureau (three years),[4] and special courses at each local railway bureau (four to eight months). This training was provided to workers who had completed higher elementary school (two years for those who completed six years

[4]JNR had six local railway bureaus: Sapporo in Hokkaido, Sendai, Tokyo, Nagoya, Kobe in Honshu and Moji in Kyushu from 1920.

Table 2.2 Technical assistants of rolling stock division of manufacturing bureau (on August 15, 1934)

Name	Monthly salary (yen)	1941 (yen)
Eikichi Yoneoka	75	
Manji Satō	70	Assitant engineer (fourth grade)
Mitsuyuki Fukazawa	70	〃
Suehito Tanaka	70	〃
Masao Yamazaki	70	〃
Masami Morita	69	
Shunpei Imoo	65	
Kōzō Noguchi	65	
Kiyoshi Nemoto	62	Assitant engineer (fourth grade)
Shunichi Ōshima	61	Assitant engineer (fifth grade)
Toyokazu Maki	61	〃
Eiichi Saiga	61	
Hitoshi Honda	61	〃
Toshio Yamamoto	61	
Teijiro Wakabayashi	60	
Isamu Fujita	59	〃
Sakuo Motoyama	59	〃
Masao Hosokawa	58	
Daizō Hanada	58	〃
Shūichi Nakajima	58	
Sadao Chine	58	
Kōtarō Uchida	58	
Kotokichi Nagamori	57	
Kō Sone	57	
Turukichi Ishii	57	〃
Kumakichi Takaoka	56	
Etsuji Tanaka	55	Assistant engineer (sixth grade)(1)
Keizō Yajima	55	Assistant engineer (sixth grade)
Yasuo Higuchi	55	〃
Gorō Ōsawa	55	
Isamu Nagashima	55	〃
Yasuo Hayashi	54	〃
Asagorō Enomoto	54	〃
Yukio Watanabe	53	〃
Ichiro Inoue	52	〃
Zensaku Yoshizawa	51	74
Tosaburō Noda	51	Assistant engineer (sixth grade)
Shigetaka Sekiguchi	51	72
Kō Honda	51	

(continued)

Table 2.2 (continued)

Name	Monthly salary (yen)	1941 (yen)
Gorō Tōya	50	74
Kikue Matsumoto	50	
Yoshitaka Nagai	49	
Tetsumi Suzuki	49	
Sôsaku Okada	48	
Tei Senda	48	69
Katsutarō Morikawa	47	
Sadakichi Tanaka	47	
Masao Kudo	46	67
Yôji Takaoka	1.42(daily)	
Kiyoharu Iijima	1.35(daily)	68
Tōichi Mochizuki	1.27(daily)	
Iwao Okamatsu	1.23(daily)	

Source Tetsudō Daijinkanbō Jinjika (1934 and 1941)
Note (1) Factory division of manufacturing bureau

elementary schools). Those who completed in-house training at each local railway bureau could continue their education, selecting among two-year routine and advanced courses at the Ministry of Railways. Workers who entered in-house training at local railway bureaus and of the Ministry of Railways were upgraded to posts formerly held by graduates of middle school and technical colleges although they were obligated to work twice as long as their period of education (Mikami 2004, 225).

The ascent of workers to assistant engineers or engineers through in-house training diffused throughout large Japanese private companies and national enterprises around WWI. The Naval Training School for Assistant Engineers (*Kaigun Gite Yôseijo*) was equivalent to JNR's *Tetsudō Kyōshūjo*. It was established at Yokosuka Naval Shipyard in 1919; subsequently, it moved to Kure Naval Shipyard in 1928. Open to talented workers who passed stringent examinations, its three-year curriculum was equivalent to technical college. Initially, comprising two divisions—shipbuilding and marine engines—it added a division of weaponry after redeployment to Kure. To qualify for entrance examinations, workers had to be 21 years old with at least three years of experience in the workshop. In prewar Japan, the division between engineers and workers was sharp and based on academic careers. In-house training provided by JNR and the navy was a means whereby ambitious workers could advance as assistant engineers and engineers. It also opened channels through which JNR and the navy could recruit engineers with practical workshop experience and knowledge of shop culture. It was sure that the existence of this carrier pass encouraged workers to upgrade their posts and experiences.

As mentioned, the Imperial Japanese Navy employed two kinds of engineers in shipbuilding: naval technical officers and naval engineers. The former were university and technical college graduates who took a military training after

Table 2.3 The number of shipbuilding officers (naval officers and engineers) at the end of Pacific War (person)

By placement	Officers/engineers	Graduates of technical colleges and universities	Graduates of naval training school
Naval ships headquarter	33	33	
	24	6	13
Technical institute of the navy	11	11	
	11	4	4
Yokosuka naval shipyard	60	60	
	19	5	11
Kure naval shipyard	65	65	
	31	11	18
Sasebo naval shipyard	51	51	
	15	1	11
Maizuru naval shipyard	42	42	
	12	5	4
Naval ships headquarter Inspectors of shipbuilding	41	40	
	44	8	34
Manufacturing bureau of the navy Others	41	41	
	9	1	8
Naval air headquarter	3	3	
	3	2	
Ministry of munition	2	2	
Ministry of the navy	1		
Total	350	348	
	168	43	103

Source Kaigun Zōsenkai Kaiin Gyōseki Kenshō Shiryō Sakusei Iinkai (1988), pp. 10–27
Note The upper is naval officers, and the lower is naval engineers

employment; the latter were graduates of universities and technical colleges who continued to serve as engineers. Naval technical officers potentially could be promoted to vice admiral. Table 2.3 surveys the composition of the Japanese navy's shipbuilding officers (technical officers and engineers). At the end of the Pacific War, 348 of 350 naval technical officers were university and technical college graduates, 43 of 168 naval engineers were university and technical college graduates, and 103 were graduates of the naval training school for assistant engineers. In the case of the fourth division of the naval headquarter in charge of designing of naval ships, the number of graduates of the naval training school who became assistant engineers were much larger than university and technical college graduates who became naval engineers. This was also the case at Yokosuka, Kure, and Sasebo naval shipyards. Table 2.3 confirms the significance of engineers who began as shop workers at the naval shipyards. Engineers who had graduated from universities and technical colleges could not manage workshops without help from assistant engineers promoted from among workers. Graduates of JNR's and the navy's training schools were bilingual in "shop culture" and "school culture".

Entering the naval training school for assistant engineers was a narrow path. Among the tens of thousands of shipyard workers, only 20 to 40 graduated from the program each year, although their numbers ballooned during the war. However, naval training school was an important path for workers to become assistant engineers or engineers, especially if they could not afford higher technical education. Such was also the case at JNR's *Tetsudō Kyōshūjo*.

2.2.2 JNR Research Institute of Railway Technology

The Ministry of Railways established a research institute in 1907 (its name was changed and its curriculum upgraded in 1910, 1913, 1920, and 1942). As Table 2.1 shows, it had five sections in 1934. There were 24 engineers, including section chiefs, for 23 of whom academic careers could be confirmed: 14 were graduates of Tokyo Imperial University, six were graduates of Tohoku Imperial University, two were graduates of Kyushu Imperial University, and one was a graduate of Kyoto Imperial University.

Expenses for the research institute, excluding salaries of engineers and assistant engineers, were around 400,000 yen from 1927 to 1938, rising to 740,000 yen in 1940 and 1,070,000 yen in 1942 (Sawai 2015, 51–52). The institute employed 294 staff in 1935, 496 in 1942, and 1,567 at the end of October 1947, when it received many ex-military and aeronautics engineers unemployed after the war (Tetsudō Gijutsu Kenkyūsho 1957, 41–42).

2.2.3 Manufacturing Division of South Manchuria Railway Company and Its Research Institute of Railway Technology

As Table 2.4 shows, the manufacturing division of South Manchuria Railway Company (SMR) had four sections; planning, locomotives, rolling stock, and machinery. The manufacturing division included a chief (Hidetsugu Nonaka, graduate of Tokyo Imperial University), four section chiefs, and 48 assistant engineers. Of the four section chiefs, three were graduates of Lushun Institute of Technology (LIT) established in 1910 and upgraded to university status in 1922. The chief of the machinery section graduated from Kumamoto Technical College. Among the 29 assistant engineers whose academic careers could be confirmed, 11 graduated from South Manchuria Technical College (SMTC) , nine from LIT, four from a technical college in Japan, two from Kyushu Imperial University, and three

Table 2.4 Engineers of manufacturing division of South Manchuria Railway (on the 1st September, 1934)

Post	Name	Affiliated colleges and universities department	Major	Year of graduation
Division chief/engineer	Hidetsugu Nonaka	Tokyo Imperial University, Engineering	Machinery	1913
Section chief of planning/engineer	Masatsugu Kubota	Lushun Institute of Technology	〃	1914
Section chief of locomotives/engineer	Shintarō Yoshino	〃	〃	1918
Section chief of rolling stock/engineer	Yoshizumi Ichihara	〃	〃	1916
Section chief of machinery/engineer	Shigeru Ishio	Kumamoto Technical College	〃	1911
Assistant engineer	Tatsuya Azuma	Yonezawa Technical College	〃	1932
〃	Hiroshi Kojima	Lushun Institute of Technology	〃	1921
〃	Yōtarō Ichijō	〃	〃	1922
〃	Michiji Fukui	〃	〃	1924
〃	Ryōichi Morimoto			
〃	Kōjirō Kawaguchi	〃	〃	1924
〃	Itaro Horiuchi	Kyushu Imperial University, Engineering	〃	1927
〃	Taneo Saeki	Lushun Institute of Technology	〃	1922
〃	Tatsuji Hidaka			
〃	Takeo Hirayama			
〃	Kiyoaki Yanagishita	Kyushu Imperial University, Engineering	〃	1929
〃	Yoshio Okamoto			
〃	Takeo Namiki	Lushun Institute of Technology	〃	1925
〃	Tatsuzō Kumioka			
〃	Azuma Yamaguchi	South Manchurian Technical College	〃	1928
〃	Tetsuya Morita	Waseda University, Science and Engineering	〃	1928
〃	Shigeo Fukushima			
〃	Washirō Nakamura	South Manchurian Technical College	〃	1917
〃	Chikao Sagawa	Lushun Institute of Technology	〃	1929

(continued)

Table 2.4 (continued)

Post	Name	Affiliated colleges and universities department	Major	Year of graduation
〃	Kazuo Tayama	〃	〃	1931
〃	Sadayoshi Hosoda	South Manchurian Technical College	〃	1927
〃	Tatsuo Maejima	〃	〃	1928
〃	Yoshihiko Higo			
〃	Shōji Sano			
〃	Genichi Etō	〃	〃	1928
〃	Kazutoki Yoshida			
〃	Masao Sone			
〃	Toshien Song	Nihon Daigaku Technical College	〃	1927
〃	Mitsuyoshi Ishihara			
〃	Shiro Hidejima	Lushun Institute of Technology	〃	1933
〃	Hachiro Inamoto	Fukui Technical College	〃	1929
〃	Masao Kawai	Yokohama Technical College	〃	1927
〃	Tsurukazu Mikami	South Manchurian Technical College	〃	1930
〃	Masuo Mizoo	〃	〃	1930
〃	Yasuzō Koga	〃	〃	1931
〃	Toshio Kazumaru			
〃	Tatsuichi Ōkubo			
〃	Toshio Tokawa			
〃	Kishirō Kumagai	〃	〃	1932
〃	Yoshio Minemura			
〃	Tamaji Nishiyama	〃	〃	1933
〃	Kazuo Katō	Nihon Daigaku, Special Courses	〃	1923
〃	Hiroshi Takahama	Fukui Technical College	〃	1933
〃	Yukiyoshi Murakami			
〃	Masatsugu Tanigawa			
〃	Susumu Katō	South Manchurian Technical College	〃	1929
〃	Toshio Samejima			
〃	Yasuna Takeuchi			

Source Minami Manshū Tetsudō Kabushiki Kaisha (1934), and Nikkan Kōgyō Shinbunsha (1934) from other schools. These figures confirm the intimate ties among SMR, LIT, and SMTC in supplying engineers.

LIT had four departments—machinery, electricity, mining, and metallurgy—and added a department of applied chemistry and aeronautics during the war. Upon its elevation to a university in 1922, the Agency of the Kwantung Leased Territory indicated the following (Kantōchō (Agency of Kwantung Leased Territory) 1922).

> More than 10 years ago young Chinese students were eager to go to Japan for study. However Japanese people took a short view, did not keep good relationships with Chinese students, often lacerated them with the result of emergence of anti-Japanese sentiment among them. […] the place of south Manchuria is the paradise for them. The Chinese students of LIT and South Manchuria Medical School in Lushun enjoy campus life in a good relationships with Japanese students. […] if we could turn out people who would become leaders of China in the future, that would largely contribute to our national interest.

Although LIT was established in China, there were only 98 Chinese among its 1,333 graduates in 1937 (Sawai 2012, 25).

SMR established South Manchuria Technical School in Dalian in 1911. Its five sections—civil engineering, architecture, machinery, electricity, and mining—covered four years of schooling. SMTC succeeded it in May 1922.

Shintarō Yoshino, chief of the locomotive section of the manufacturing division, graduated from LIT's machinery department in 1918 and began his career as an engineer at SMR's Shahekou railway factory in Dalian. He studied at the American Locomotive Company, one of the world's largest locomotive builders, from 1923 to 1925. He designed the fastest locomotive in the Japanese empire and Manchuria, which pulled streamlined passenger cars called "Asia," in 1933 (Sawai 2015, 6–7). According to Yoshino, SMR imitated, then improved the imported locomotives, mostly made in the US, for about 20 years after establishment. In 1927, a Pashiko-type locomotive was for the first time designed based not on imported locomotives but on original ideas from SMR engineers (Yoshino 1939, 36).

The SMR also established the research institute of railway technology in 1922 on the advice of Yasujirō Shima, who retired from JNR in 1919 and became a director of SMR. Table 2.5 notes 44 primary researchers whose academic careers could be confirmed among 52 posted as chief researchers from 1923 to 1942. Seven were LIT graduates, and three were SMTC graduates. This proportion is relatively small compared with that in the manufacturing division, in which 20 were graduates of imperial universities and 15 graduated from technical colleges in Japan. However, engineers frequently transferred among SMR's railway factories, manufacturing division, and research institute.

Many SMR researchers had studied abroad (Table 2.5). In 1911, SMR established rules that prescribed two kinds of foreign study: two-year study abroad, especially in the US and Europe, for those with five or six years of experience and study abroad for officer candidates with 10 years of experience at SMR. SMR often sent engineers and researchers on six-month sojourns to study advanced

Table 2.5 Main researchers of South Manchuria Railway Company and the research institute of the SMR

Name	Name of post	Affiliated colleges and universities department	Major	Year of graduation	Remarks
Kiyoshi Takemura	Interlocking director (23)	Tokyo Imperial University, Engineering	Machinery	1902	
Aihito Inoue	Section chief of machinery & rolling stock (23), section chief of machinery (24), laboratory chief of machinery (32), assistant director of machinery department (38), assistant director of metal department (40)	Osaka Technical College	〃	1911	Official trip to Europe and USA (34)/ 8 months
Kaneyuki Ochiai	Section chief of electricity (23)	Lushun Institute of Technology	Electricity	1915	
Kingo Kaise	Interlocking director (24)	Kyoto Imperial University, Engineering	Machinery	1901	
Yoshio Nakahara	Section chief of rolling stock (25)				
Masatoshi Koreyasu	Section chief of rolling stock (26)	Lushun Institute of Technology	〃	1918	Study in USA (23)/ 2 years
Tadao Miyajima	Section chief of civil engineering (26), laboratory chief of civil engineering and construction (32), of water and sewerage (32)	Kyushu Imperial University, Engineering	Civil engineering	1917	Official trip to Europe and USA (36)/ 8 months
Inosuke Watanabe	Director (27), research section chief of machinery and electricity (31), of machinery (32), director (37)	Kyoto Imperial University, Engineering	Machinery	1913	
Kazuharu Kusaka	Section chief of analysis (27), research section chief of metals (34)	Lushun Institute of Technology	Mining and metallurgy	1921	Study in Europe and USA/2 years
Kojirō Takagi	Section chief of rail tracks (28), research section chief of civil engineering (31), assitant director of civil engineering depaertment (38)	Kumamoto Technical College	Civil engineering	1916	

(continued)

Table 2.5 (continued)

Name	Name of post	Affiliated colleges and universities department	Major	Year of graduation	Remarks
Teiji Nebashi	Director (31, 34), temporary director (39)				
Kanji Kurihara	Director (31)	Tokyo Imperial University, Engineering	Applied chemistry	1905	
Hiro Yamamoto	Research section chief of electricity (31)	Lushun Institute of Technology	Electricity	1922	
Matsunosuke Iwatake	Research section deputy chief of electricity (32), research section chief of light electricity (32), of heavy electricity (34) Research section chief of electricity (34), assitant director of investigation and electricity departments (38)	,	Machinery	1921	
Masuzō Isokane	Research section chief of rolling stock (32), of locomotives (35), assitant director of rolling stock department (38)	Lushun Institute of Technology	Machinery	1919	Official trip to Europe and USA (35)/ 5 months
Shōichi Fujii	Research section chief of heavy electricity (32)	〃	Electricity	1923	
Shikimatsu Tanaka	Laboratory chief of purchasing goods (32), of electrical goods (35)	South Manchurian Technical College	〃	1927	
Nobuhiro Ishibashi	Research section chief of rail tracks (32)	Kyoto Imperial University, Engineering	Civil engineering	1927	
Teiji Node	Research section chief of heat economy (33), assistant director of machinery department (40), Research section chief of machinery (40) Research section chief of heat (40), chief of design roon (40)	Tokyo Imperial University, Engineering	Machinery	1927	Official trip to Europe and USA (36)/ 6 months

(continued)

Table 2.5 (continued)

Name	Name of post	Affiliated colleges and universities department	Major	Year of graduation	Remarks
Chūji Fuse	Research section chief of civil engineering and construction (33), research section chief of structure and construction (40)	Kyoto Imperial University, Engineering	Architecture	1925	
Hanzō Omi	Research section chief of light electricity (34)	Tokyo Technical College	Electricity	1923	
Kazuo Endō	Section chief of metal analysis (35), of analysis (39), research section chief of chemistry (40)	〃	Ceramic engineering	1913	
Tokijirō Takeuchi	Deputy chief of manufacturing Sect. (35), section chief of manufacturing (39)				
Shigeru Ishio	Research section chief of rolling stock (35), assistant director of rolling stock department (38)	Kumamoto Technical College	Machinery	1911	
Itaru Kawada	Research section chief of general rolling stock (35)	Kyoto Imperial University, Engineering	〃	1927	
Ichirō Kanō	Research section chief of heavy electricity (35), of electric power (38)	Tohoku Imperial University, Engineering	Electricity	1930	
Shigeyoshi Kutsukake	Research section chief of rail tracks (35)	Nagoya Technical College	Civil engineering	1924	
Takuitsu Kamisaka	Research section chief of machinery (35)	Kyushu Imperial University, Engineering	Machinery	1928	Study in Europe and USA (33)/ 2 years
Tsuneya Marusawa	Director (36)	Kyoto Imperial University, Engineering	Applied chemistry	1907	
Kiyoaki Yanagishita	Research section chief of general rolling stock (36)	Kyushu Imperial University, Engineering	Machinery	1929	

(continued)

Table 2.5 (continued)

Name	Name of post	Affiliated colleges and universities department	Major	Year of graduation	Remarks
Tan Furuno	Research section chief of light electricity (36), of cable communication (38), of general communication (40)	Tohoku Imperial University, Engineering	Electricity	1930	
Toshitarō Matsunawa	Assistant director of standard investigation (38)	Lushun Institute of Technology	〃	1917	
Yoshihiko Hara	Research section chief of metals (38)	Tohoku Imperial University, Engineering	Metal	1929	
Hideo Itatani	Research section chief of wireless communication (38), of cable communication (40)	〃	Electricity	1934	Study in UK and Germany (36)/2 years
Kikuo Matsui	Research section chief of signals (38)	Kobe Technical College	〃	1926	
Shûsaku Kamiya	Research section chief of rail tracks (38)	Tokyo Imperial University, Engineering	Civil engineering	1932	
Sueyoshi Kojirō	Research section chief of bridges (38), of steel structure (40)	Kumamoto Technical College	〃		
Hiroshi Yamaoka	Chief of standard investigation (38), of standard investigation and architecture group (39)	South Manchurian Technical College	Architecture	1925	
Toyozō Dōke	Research section chief of locomotives (39)	Tokyo Technical College	Machinery	1929	
Hajime Hirose	Research section chief of wireless communication (39)	Yonezawa Technical College	Electricity	1931	
Fusatarō Aizawa	Section of standard investigation, civil engineering group (39)				
Isamu Satō	Research section chief of machinery (39)	Hokkaido Imperial University, Engineering		1932	

(continued)

Table 2.5 (continued)

Name	Name of post	Affiliated colleges and universities department	Major	Year of graduation	Remarks
Tsuyoshi Oka	Assistant director of internal combustion department (40)	Tohoku Imperial University, Engineering	Machinery	1927	
Eizaburō Yamauchi	Research section chief of automobiles (40), of mechanism of automobiles (41)				
Shōju Kawarai	Research section chief of wireless communication (40)	Sendai Technical College	Electricity	1933	
Toshiaki Takabayashi	Research section chief of concrete (40)				
Yu Onogawa	Research section chief of structural welding (40)	Kumamoto Technical College	Civil engineering	1932	
Morikazu Takahashi	Research section chief of civil engineering (40)	Nagoya Technical College	〃	1919	
Keigo Yoshizaki	Section chief of electrical goods experiments (40)	South Manchurian Technical College	Electricity	1933	
Manrei Yoneda	Research section chief of internal combustion (41)				
Rinzō Asahi	Section chief of metal analysis (41)	Osaka Technical College	Applied chemistry	1914	
Sōichi Nishikawa	Director (42)	Kyoto Imperial University, Engineering	Civil engineering	1916	

Source Nikkan Kōgyō Shinbunsha (1934), Gakushikai (each year), Mimami Manshū Tetsudō Kabushiki Kaisha Tetsudō Gijutsu Kenkyūsho (1943), pp. 103–116, Mantetsu Tekkenkai (1990), pp. 153–161
Notes (1) Figures in parentheses of name of post are the first year of placement
(2) Figures in parentheses of remarks are departure year of official trip and study abroad

locomotives and rolling stock and participate in international academic conferences (Amano 2012, 22–23).

For engineers in prewar Japan, study abroad and overseas business trips were important for building skills. For example, Mitsubishi Electric Co. Ltd., which had technological ties with US-based Westinghouse in 1923, sent more than 60 engineers abroad before 1939 (Sawai 2015, 36–37).

2.2.4 *Colonial Government Railways: Taiwan and Korea*

During the first half of the 1930s, Taiwan stopped importing steam locomotives from Japan, and the Taipei and Gaoxiong factories of the colonial government railways began to manufacture rolling stock for passengers and freight (Sawai 2015, 58).

As Table 2.6 shows, 27 engineers and assistant engineers worked in the manufacturing section of the railway division of the transportation bureau of Taiwan Sōtokufu (TS) in 1934. Among the 16 engineers and assistant engineers whose academic credentials could be confirmed, one was an LIT graduate and one an SMTC graduate.

In Korea, new rolling stock as well as steam locomotives were manufactured in Keijo (Seoul) and Pusan at factories of colonial government railways. Colonial government railway bought large quantities of rolling stock from designated private companies besides manufacturing it in-house.

As Table 2.7 shows, six engineers from the manufacturing division of the railway bureau of Chōsen Sōtokufu (CS) in 1934 were graduates of Tokyo Imperial University, Kyushu Imperial University, and Tokyo Technical College. Engineers at railway factories included graduates of LIT, and we could confirm that many graduates of LIT and SMTC were assistant engineers in railway factories.

Table 2.8 lists *Tetsudō-shu* (junior assistant engineers, a post established during a 1925 ministerial reorganization) (Chong 2008, 496) in 1934. They supported engineers and assistant engineers in workshops. Promotion to junior assistant engineer required that workers had at least five years of workshop experience and satisfactory performance (Chong 2008, 503). As an in-house training facility for employees of the railway bureau of the CS, SMR who managed government railways in Korea at that time, established the Seoul Railway School in 1919. Although the school was abolished in March 1925, another superseded it the following month. It included four courses: the regular, manufacturing, telegraph, and short. The regular course featured three sections: management, civil engineering, and driving. Admission qualifications for the regular course were completion of three years of higher elementary school (14 years old). Admission to the four-year manufacturing course (for factory workers) required completing elementary school (12 years old). The admission and schooling requirements were altered to completion of higher elementary school and two years of schooling in March 1928. After admission, trainees in the manufacturing course agreed to be apprenticed to workshops and received allowances (Railway Bureau of Governor-General of Korea 1934, 87–88). From 1925 to 1941, it graduated 269 students: 118 Koreans and 151 Japanese (Chong 2008, 508).

As an example of engineers who worked at the railway bureau of CS, we follow Kiichi Imai (1912–2005) in prewar and postwar periods. After graduating from Komagome Middle School in Tokyo, he completed the Concrete Vocational School (established in 1929 with a two-year curriculum) in 1934. He started his career as an engineer at the CS railway bureau in 1934, was promoted to assistant engineer in

Table 2.6 Engineers of section of manufacturing of railway division of transportation bureau of Governor-General of Taiwan (on August 1st, 1934)

	Name	Affiliated colleges and universities department	Major	Year of graduation
Engineers	Kazuhiko Hayami	Kyoto Imperial University, Engineering	Machinery	1913
	Shunichi Azuma	Kyushu Imperial University, Engineering	〃	1925
	An Ogawa	Sendai Technical College	〃	1917
Assistant engineers	Tadao Yamazaki	South Manchurian Technical College	Architecture	1929
	Kiichi Itō			
	Misoji Watanabe			
	Tatsuji Tsuyuki	Hokkaido Imperial University, Engineering		1928
	Itoku Ōmachi	Nagoya Technical College	Machinery	1927
	Shichitarō Okamoto	Osaka Technical College	〃	1920
	Katsushi Iwaki			
	Kiichi Nakajima	Sendai Technical College	〃	1922
	Kōichi Koizumi			
	Yasuyoshi Hirakawa	Lushun Institute of Technology	〃	1930
	Teiji Igawa			
	Ushichi Ōmura	Waseda University, Science and Engineering	〃	1925
	Wataru Satō	Kanazawa Technical College	〃	1925
	Takashi Shingō			
	Mitsuyoshi Miwa			
	Keiichi Honda	Nagoya Technical College	〃	1925
	Akira Yoshino	Kanazawa Technical College	〃	1926
	Kei Satō	Waseda University, Science and Engineering	〃	1929
	Kenji Yasuda			
	Yoshio Katō	Nagaoka Technical College	〃	1927
	Kiyoshi Shimizu			
	Seiji Motoyoshi			
	Toshio Matsuno	Kumamoto Technical College	〃	1928
	Hiroshi Sakata			

Source Taiwan Sōtokufu (1934), and Nikkan Kōgyō Shinbunsha (1934)

the construction division, worked at the CS research institute of government railways, and retired from there in 1946. After returning to Japan, he worked part-time and thereafter full-time at JNR's research institute in May 1947. He founded MARUTO Testing Machine Co., Ltd. in December 1951 after retiring from the

Table 2.7 Engineers of manufacturing division of railway bureau and railway factories of Governor-General of Korea (on July 1st, 1934)

	Name	Affiliated colleges and universities department	Major	Year of graduation
Engineers of the manufacturing division of railway bureau	Sadaharu Fukumi	Tokyo Imperial University, Engineering	Machinery	1915
	Hajime Kan	〃	〃	1919
	Sanichirō Sakiyama	〃	〃	1922
	Yoshimitsu Hara	Tokyo Technical College	〃	1915
	Shūji Imai	Kyushu Imperial University, Engineering	〃	1920
	Shigeharu Toda	Tokyo Technical College	〃	1919
Assistant engineers of the manufacturing division of railway bureau	Masanobu Sugimoto	Lushun Institute of Technology	〃	1920
	Rikio Tsunatani	Tokyo Technical College	〃	1921
	Ichiro Tsuchisaki	Yonezawa Technical College	〃	1921
	Hiroshi Maeda	Lushun Institute of Technology	〃	1921
	Masao Somatani	Nagoya Technical College	〃	1921
	Ryōzō Yonezawa	Tokyo Technical College	Electricity	1922
	Shigemichi Nishiyama	〃	Machinery	1923
	Tsuruzō Hongō	Kyoto Technical College	Design	1923
	Nikichi Fujii			
	Tetsuichi Yamada			
	Young-jik Kang	Waseda University, Science and Engineering	Electricity	1927
	Tōru Yoshimori			
	Kiichi Satō			
	Eung-gil Lee	Osaka Technical College	Machinery	1926
	Taku Ishii			
	Takashi Wada	Tokyo Institute of technology	Machinery	1932
	Naoya Maehara	Lushun Institute of Technology	〃	1925
	Nobukichi Satō	Sendai Technical College	〃	1926
	Kaizō Furuya	Waseda University, Science and Engineering	〃	1926
	Manpei Shimizu			
	Rizō Mitsutani	Keijo (Seoul) Technical College	Applied chemistry	1921
	No-su Kim	Waseda University, Science and Engineering	Machinery	1929

(continued)

Table 2.7 (continued)

	Name	Affiliated colleges and universities department	Major	Year of graduation
	Takeo Tanaka			
	Tatsuo Iwasaki	South Manchurian Technical College	Machinery	1923
	Li-hyeong Chon	Hamamatsu Technical College	Electricity	1927
	Kenichi Fukatsu	〃	Machinery	1928
	Myon-gyu Chang			
	Mong-sik Gwon	Yonezawa Technical College	Machinery	1929
	Ei Fukuya	Hiroshima technical College	〃	1929
	Mitsuo Saitō	Meiji Technical College	Applied chemistry	1927
	Torao Nakajima			
	Kiyoshi Ichinose	Lushun Institute of Technology	Electricity	1930
	Tatsusaburō Ikuta			
	Tōru Fujii	Kyoto Imperial University, Engineering	Machinery	1931
	Sumio Satō			
	Jirō Hosoya	South Manchurian Technical College	Machinery	1931
Pusan factory engineers	Sakae Hayashibe	Osaka Technical College	Machinery	1912
	Heiichi Noguchi	Tokyo Imperial University, Engineering	〃	1927
Pusan factory assistant engineers	Hidetsuna Sasaki	Waseda University, Science and Engineering	〃	1917
	Michio Higuchi	〃	〃	1927
	Takasaku Uzawa	Yamanashi Technical College	〃	1928
	Torao Matsui	Sendai Technical College	Electricity	1928
	Minoru Shiode			
	Ikuzō Fujinawa	Kyushu Imperial University, Engineering	Machinery	1930
Seoul factory engineers	Hidesuke Miyahara	Tokyo Imperial University, Engineering	〃	1917
	San Aoki	Osaka Technical College	〃	1913
	Nobukazu Satō	Lushun Institute of Technology	Electricity	1915
	Hōkichi Aihara	Tokyo Imperial University, Engineering	Machinery	1926
	Mankichi Kawamoto	Lushun Institute of Technology	〃	1919
	Sanshirō Kawashima	Tohoku Imperial University, Engineering	〃	1927

(continued)

Table 2.7 (continued)

	Name	Affiliated colleges and universities department	Major	Year of graduation
Seoul factory assistant engineers	Sanji Igarashi			
	Kaku Iwasaki	Lushun Institute of Technology	Machinery	1920
	Kenzō Takahashi	〃	〃	1921
	Yoshiyuki Ōno			
	Yasushi Tazawa	Tokyo Technical College	Machinery	1922
	Yasuji Nagayasu	Meiji Technical College	〃	1923
	Shirō Yoshida	Tokyo Technical College	〃	1924
	Jirō Miyazono	Kyushu Imperial University, Engineering	〃	1926
	Shigeo Inoue			
	Osamu Ezaki			
	Taneo Harada			
	Toshio Higuchi	Meiji Technical College	Machinery	1924
	Yoshimasa Tokitō	Hokkaido Imperial University, Engineering		1929
	Fusakichi Saitō			
	Yoshio Yamada	Lushun Institute of Technology	Machinery	1929
	Tamotsu Nyutani			
	Nakahiko Yasuki			
	Kamejirō Izawa	Lushun Institute of Technology	Machinery	1923
	Yoshio Ioribara			
	Takeo Tsuchikawa			
	Kiyonobu Aoki	Hamamatsu Technical College	Machinery	1927
	Haruzō Maeda	South Manchurian Technical College	Mining & Machinery	1927
	Hiroo Miyoshi			
	Setsuzō Kobayashi	Yonezawa Technical College	Machinery	1927
	Fumio Satō	Kumamoto Technical College	〃	1928
	Kusue Kawachi	Hiroshima Technical College	Machinery, special courses	
	Kazuo Nojima	Meiji Technical College	Machinery	1928
	Kazuo Sakurai	Hamamatsu Technical College	Electricity	1929

(continued)

Table 2.7 (continued)

	Name	Affiliated colleges and universities department	Major	Year of graduation
	Kiyonojō Nakano			
	Tsukuo Aoki			
	Toshio Shiba			
	Chōshichi Kawamura	Hiroshima Technical College	Machinery	1928
	Kazuo Tashiro	Yokohama Technical College	〃	1930
Pyongyang branch factory assistant engineers	Goshichi Miki			
	Shunichi Koga			
	Hakuai Ōsaki	Waseda University, Science and Engineering	Machinery	1924
	Masuo Kanzaki	South Manchurian Technical College	〃	1923
Chongjin factory engineers	Takesaburō Miyauchi	Tokyo Technical College	〃	1910
Chongjin factory assistant engineers	Hanji Tominaga			
	Kyūhei Tsuboi			
	Naokichi Inaba	Waseda University, Science and Engineering	Electricity	1928

Source Chōsen Sōtokufu (1934), Chōsen Kōgyō Kyōkai (1939), and Nikkan Kōgyō Shinbunsha (1934)
Notes (1) The blank is unknown
(2) The first line of engineers and assistant engineers of each railway factory is a factory superintendent

Table 2.8 Tetsudō-shu (junior assistant engineers) of Keijo (Seoul) factory of railway bureau of Governor-General of Korea (on July 1st, 1934)

Name	Monthly salary (yen)
Sun-dal Kim	98
Ichigorō Arai	95
Yutaka Hano	95
Kiyoharu Shimoura	93
Ushitaro Igarashi	90
Tomokichi Yamada	90
Makoto Tanaka	89
Takisaburō Sekiya	88
Jon-tae Kim	88
Hideharu Miyawaki	88
Hanjūrō Tsukamoto	86
Yon-il Choe	84
Matsutarō Ōnishi	84
Ryūnoshin Matsumiya	82
Sadao Yoshimune	81
Kisuke Yoshimoto	80
Shigetsugu Ishiguro	77
Mito Sasaki	77
Morifumi Arima	77
Tatsuzō Fukuda	76
Takaichi Ōno	76
Tomesaku Kinoshita	76
Kinsaku Kurakata	74
Otojirō Komatsu	73
Tomotsuchi Ōyama	66

Source Chōsen Sōtokufu (1934)

institute in 1949. In 1953, Imai developed Japan's first instrument to measure air in ready-mixed concrete.[5]

Dalian Machine Works (*Dairen Kikai Seisakusho*) employed15 engineers in 1934, 12 of them graduated from LIT and SMTC and three from technical colleges in Tokyo, Yokohama, and Nagoya. At Yongsan Manufacturing Co., Ltd. (*Ryūsan Kōsaku*) in Seoul, a representative maker of rolling stock in Korea, there were four LIT graduates among 11 engineers in 1934 (tallied from Nikkan Kōgyō Shinbunsha 1934).

As above-mentioned, LIT and SMTC played a great role in supplying engineers for colonial government railways in Taiwan and Korea, SMR, and rolling stock and locomotive companies in Dalian and Seoul. By 1942, the number of living LIT

[5]HP of MARUTO Testing Machine Company.

graduates approached 1,337, of which 660 were working in "Manchukuo" and the Kwantung Leased Territory, 130 in China, 76 in Korea, 24 in Taiwan, and 436 in Japan (Sawai 2012, 28).

2.3 Collaborative Research and Institutionalized Interactions Within Organizations

2.3.1 Conference on Rolling Stock

Among the technological activities led by JNR, the Conference on Rolling Stock (*Sharyō Kenkyūkai*) was significant. The conference was established in 1922 as an annual meeting, becoming biannual after 1925. Through 1942, 19 conferences addressed locomotives and 17 concerned freight cars and passenger cars. Other topics of common interest were raised, sponsored by JNR's manufacturing bureau (Kubota 1981, 107).

The group of JNR engineers from the rolling stock division of the manufacturing bureau formed core of *Sharyō Kenkyūkai*, augmented by engineers from SMR, government railways in Taiwan and Korea, private railway companies, materials and parts producers, and designated factories. In this sense, the network of collaborative research organized by the rolling stock division of JNR's manufacturing bureau spread throughout the Japanese Empire.

Activities of the conference, where 70 to 100 engineers gathered, exceeded the few days of sessions. Conference topics were announced a year in advance so that members could conduct related research in their own workshops and research institutes beforehand (Tetsudō-shō Kōsakukyoku 1925, 16). Results were discussed at the conference. JNR, SMR, and government railways in Taiwan and Korea promptly put some into practice (Nihon Kokuyū Tetsudō (JNR) 1972, 446–453). Thus, the conference, where collaborative research could be evaluated via exchanges between users and builders of rolling stock and locomotives, contributed to the technological upgrading of the Japanese empire's rolling stock industry.

Okiie Yamashita, former chief of JNR's manufacturing bureau, noted (Inoue 1952, 241).

> As many difficult research themes were delivered to the Conference on Rolling Stock, the conference often could not discover the best solution in a short time, carried discussion forward in a few years. The research institute of JNR played an important role in solving the problems, with the result of upgrading of the presence of the research institute in JNR.

Practical research topics covered by the Conference on Rolling Stock gave direction to the research institute, minimizing research for its own sake and promoting research for practical application.

2.3.2 Technical Committee on Rolling Stock

Kiichi Asakura, former chief of JNR's manufacturing bureau, went to Manchuria with an opinion concerning the unification of materials and parts for rolling stock of JNR, SMR, and government railways in Korea, which had different gauges.[6] He gathered support from Inosuke Watanabe, chief of SMR's manufacturing division, and from Sadaharu Fukumi, chief of the manufacturing division of Korea's government railway, in 1938.

The Technical Committee on Rolling Stock was established in July 1939 and held five large-scale meetings from November 1939 to December 1941(Nihon Kokuyū Tetsudō (JNR) 1958, 147). As Table 2.9 indicates, members of the committee were officials and engineers from JNR, the Ministry of Commerce and Industry, SMR, government railways in Taiwan and Korea, North China Transportation Co., Ltd., and Central China Railways Co., Ltd. Dependent on educational organizations, such as LIT and SMTC located in Dalian and Lushun with regard to the supply of engineers, railways within the Japanese empire were organized into an ellipse with JNR and SMR at its apex.

2.3.3 Study Group on Factory Practices

JNR had 22 railway factories in Japan during the war, ranging from mid-sized operations with 300 employees to large factories with 3,000 to 4,000 workers. As national factories exempt from market competition, JNR needed to pursue technological upgrading without input from the market. Among these measures we could confirm the mechanism that promoted technical interactions among the various kinds of workshops (Mukasa 1941, 96–98). At JNR, for example, a railway factory with 1,000 workers held about 10 workshops covering assembly, fitting, machine, cast iron, lathes, wood working, and painting. A factory director, engineers, assistant engineers, and foremen participated assembly. They inspected a workshop for three or four hours, then 20 or 30 persons discussed how to upgrade its efficiency.

In addition, the Study Group on Factory Practices (*Kōjō Gyōmu Kenkyūkai*) was started in 1927. Representatives of the same workshops at different factories met quarterly (Mukasa 1941, 99–104). For example, each machine workshop at a large factory was in charge of this study group once every two-and-a-half to three years. Among the activities of Study Group on Factory Practices, technical experts at each factory inspected the workshop concerned all day and discussed problems. Inspection and discussion topics included locomotives and air brakes, boilers and welding, casting, forging, lathes and other machine tools, electric machinery,

[6]JNR adopted a narrow gauge. Governmental railways in Korea and the SMR employed standard gauges.

Table 2.9 Attendants of the fourth technical committee on rolling stock (July 27, 1941)

Post	Division	Name
Chief of the sixth division of planning agency	Director of planning agency	Yaichirō Shibata
First section chief of the sixth division	Secretary	Masatake Tsutsumi
First section of the sixth division	Engineer	Sadanori Shimoyama
Third section chief of the sixth division	Investigator	Masanobu Yoshii
Third section of the sixth division	Secretary	Kunio Yoshizawa
〃	Assitant engineer	Yogorō Wakamatsu
Second section of the second division	Engineer	Michio Yamaza
〃	〃	Kazutsune Murata
Second section of the seventh division	〃	Kōhei Suzuki
Section chief of economy of the office for Manchuria		Haruji Yoshida
Economic division of Agency for development of Asia	Official	Yutaka Abiko
Technology division	Engineer	Tatsuo Kuribayashi
Transportation section of purchasing bureau of the Army	Lieutenant in the Army	Kōhei Eguchi
General affairs section of general affairs bureau of the Ministry of Commerce and Industry	Engineer	Naosaburō Torii
〃	〃	Kiyoshi Oda
First section chief of rolling stock of manufacturing bureau of the Ministry of Railways	Engineer	Kingo Mukasa
First section of rolling stock	〃	Hideo Shima
〃	〃	Ryōtarō Asamura
〃	〃	Michiaki Utashōji
First section chief of transportation of transportation bureau	〃	Nobuichirō Hamano
First section of transportation of transportation bureau	〃	Kazuo Okada
First section chief of purchasing of purchasing bureau	Official	Toshio Hata
First section of purchasing	Engineer	Gōzō Takigawa
Third section chief of railway investigation bureau	〃	Yoshitane Azuma
Third section of railway investigation bureau	〃	Ryōrō Yokoi
Section of industry of encouragement of industry bureau of the Ministry of Development	Official	Yasuhiko Toyomura

(continued)

Table 2.9 (continued)

Post	Division	Name
Section of commerce and industry of encouragement of industry bureau	Assitant	Genji Torii
Division chief of manufacturing of railway bureau of Governor-General of Korea	Engineer	San Aoki
Division of manufacturing of railway bureau	〃	Takashi Wada
〃	〃	Tōru Fujii
〃	〃	Shirō Yoshida
〃	〃	Hakuai Ōsaki
〃	〃	Naoya Maehara
Division of planning of railway bureau	〃	Shigeru Watanabe
Section chief of manufacturing of railway division of transportation bureau of Governor-General of Taiwan	〃	Kazuhiko Hayami
Transportation bureau of Agency of Sakhalin	〃	Tatsuya Murakami
Technical chief of Tokyo branch of South Manchuria Railway	Counsellor	Tomohiko Kimura
Division chief of manufacturing of manufacturing bureau of railway headquarters	Counsellor	Shintarō Yoshino
Marerial section of Tokyo branch	Vice-counsellor	Tadashi Yamada
Marerial section chief of Tokyo branch	Counsellor	Sumisaburō Kōzai
Research institute of railway technology	Vice-counsellor	Isamu Satō
Rolling stock section of trnbsportation bureau	Staff	Toshi Koizumi
Tokyo branch	〃	Yasuzō Koga
Chief of manufacturing bureau of North China Transportation	Advisor	Nakaji Katō
Mabufacturing bureau	Advisor of bureau	Yoshikazu Kagata
Technical chief of Tokyo branch	Vice-counsellor	Seiichi Miyake
Tokyo branch	〃	Kazuo Furuhara
〃	staff	Seiichirō Kaneko
Manufacturing bureau	〃	Fukujirō Okino
Transportation bureau	〃	Yoshiaki Sakato
Division chief of manufacturing of Central China Railways	Counsellor	Hajime Hirayama
Division of transportation	Vice-counsellor	Shigetake Iwasaki
Division of manufacturing	Assistant cousellor	Miyoji Yamamoto
〃	staff	Yūkichi Sakai
Technical chief of Tokyo office	Counsellor	Matsuchiyo Nomura
Kawasaki Sharyô Co., Ltd.		Bungo Shimoda

(continued)

Table 2.9 (continued)

Post	Division	Name
Kisha Seizô Co., Ltd.		Kiichi Asakura
〃		Kanichirō Hasegawa
Nippon Sharyô Seizô Co., Ltd.		Shōhachi Akiyama
Mitsubishi Heavy Industry Co., Ltd.		Mamoru Nakamoto
Hitachi Seisakusho co., Ltd.		Kiyoshi Hattori
Tanaka Sharyô Co., Ltd.		Hajime Satō
Shinuki Tekko Co., Ltd.		Eiji Asano
Teikoku Sharyô Co., Ltd.		Takasaburō Kanesaka
Sumitomo Kinzoku Co., Ltd.		Ryōkichi Kawamoto
〃		Harumitsu Suzuki
Sumitomo Kôkan Co., Ltd.		Kanichi Katō
Tokyo Kiki Co., Ltd.		Katsuyuki Nakabayashi
Mitsubishi Kôzai Co., Ltd.		Sōzō Hattori
Nippon Seitetsu Co., Ltd.		Itsuji Adachi
Rolling Stock Industry Association		Seiichi Akita
〃		Shinsuke Hashimoto

Source Tetsudō Sharyō Gijutsu Kyōgikai 1941

freight cars and wood, passenger cars and finishing, painting, by-products and stock, general affairs, and calculation (Nihon Kōgyō Kyōkai 1941, 57). The group's activities peaked around the mid-1930, just before the Japan–China War.

JNR held similar meetings with designated private factories. Its engineers inspected designated private factories and indicated areas for improvement. Each railway factory sponsored study groups. One study group named *Hatsukakai*, met monthly at JNR's Kokura factory. Around 15 members who were graduates of technical colleges or universities participated. Two members made presentations at each meeting (Kokura Kōjō of JNR 1951, 48).

2.4 Conclusion: Collaborative Research Project as a Measure for Technological Catching up

Governmental and private sector collaboration on research areas of mutual interest was commonplace in the rolling stock industry and other sectors in prewar Japan. The Conference on Machine Tools, mainly conducted by JNR's manufacturing bureau, continued from 1936 to 1940. The Conference on Wind Tunnels and Tanks, primarily sponsored by the science division of the naval aeronautics arsenal, met several times during the war. This conference facilitated collaborative research in all fields of aeronautics, involving military engineers alongside scientists from imperial universities and aircraft and engine manufacturers. Research topics were announced in advance, research results were examined, and unresolved issues were discussed at following conferences.

As noted aeronautical scientist Ichirô Tani recalled (Tani 1982, 39),

> If presenters were naval officers, there was no diffidence or flattery. This openness seemed to be specific to naval aeronautics arsenal, therefore the Conference on Wind Tunnel and Tank was much balder than lecture meetings of academic societies and much more academic than them.

Research targets were specified, and the organizers were the largest consumers of their output. Therefore, this public–private collaboration could produce great achievements in a relatively short period. However, this "efficient" collaboration also presented problems. Private sector participants largely were constrained to designated manufacturers; there was no channel through which the largest users, such as JNR and the navy, could receive results of R&D activities of non-members. Although the manufacturers were "designated" because of their relatively high capabilities, this restriction widened the divide between designated and non-designated manufacturers because only the former obtained orders and revenues from the large and stable users.

The collaborative research between users and designated manufacturers able to produce high-quality products was one aspect of the efficient R&D adopted by a latecomer to industrialization—i.e., Imperial Japan. In this method of selecting qualified manufacturers, the state—the largest buyer—determined which manufacturers were candidates for optional contracts. For non-member firms excluded from collaborative research, the closed system of R&D composed of JNR, SMR, the military, and the Ministry of Telecommunication was a kind of "crony innovation system" that has great significance for scholars of economic and business history of East Asia in the 20th century.

References

Amano, Hiroyuki. 2012. *Mantetsu tokkyū Ajia no tanjo (Advent of express "Ajia" of South Manchuria railway)*. Tokyo: Hara Shobō.

Chong, Chae-jong. 2008. *Teikoku nihon no shokuminchi shihai to kankoku tetsudō: 1892–1945 (Coronial rule of imperial Japan and railways in Korea: 1892–1945)*. Tokyo: Akashi Shoten.

Chōsen Kōgyō Kyōkai. (ed.). 1939. *Chōsen gijutsuka meibo (List of engineers in Korea)*. Chōsen Kōgyō Kyōkai: Seoul.

Chōsen Sōtokufu. (ed.). 1934. *Chōsen sōtokufu oyobi shozoku kansho shokuinroku (The list of staff of Governor-General of Chôsen and its offices)*. Seoul: Chōsen Sōtokufu.

Gakushikai. (ed.). each year. *Gakushikai shimeiroku* (The list of members of gakushikai). Tokyo: Gakushikai.

HP of MARUTO Testing Machine Company (http://www.maruto-group.co.jp/menu_1/sougyousya/)

Inoue, Tadashirō. 1952. *Kokutetsu no kaiko: senpai no taikendan (The recollection of JNR: experiences of seniors)*. Tokyo: Nihon Kokuyū Tetsudō.

Kaigun Zōsenkai Kaiin Gyōseki Kenshō Shiryō Sakusei Iinkai (ed.). 1988. *Taiheiyō sensō shūketsuji niokeru zōsenkan no haichihyō (Allocation of naval officers at the end of the Pacific War)*. Tokyo: Kaigun Zōsenkai.

Kantōchō (Agency of Kwantung Leased Territory). 1922. *Ryojun kōkadaigaku setsuritsu shuisho* (Charter of establishment of Lushun Institute of Technology). Lushun: Kantōchō.

Kokura Kōjō of JNR. (ed.). 1951. *60-nen no kaiko* (Recollection of 60 years). Kokura: Kokura Kōjō.

Kubota, Hiroshi. 1981. *Tetsurin no kiseki: tetsudō sharyō 100-nen no ayumi (The tracks of iron wheels: one hundred year history of rolling stock)*. Tokyo: Taishō Shuppan.

Mantetsu Tekkenkai (ed.). 1990. *Mantetsu Tetsudō Gijutsu Kenkyūsho Shi (The history of the research institute of the South Manchuria Railway)*. Tokyo: Mantetsu Tekkenkai.

Mikami, Atsushi. 2004. Tetsudō kyōshūjo no rekishi 2 (History of in-house training school of the JNR) In *Shokugyō to senbatsu no rekishi shakaigaku* (Historical sociology of vocation and screening), eds., Aya Yoshida and Teruyuki Hirota. Tokyo: Seori Shobō.

Minami Manshū Tetsudō Kabushiki Kaisha (SMR). (ed.). 1934. *Shainroku* (The list of staff). Dalian: Minami Manshū Tetsudō Kabushiki Kaisha.

Minami Manshū Tetsudō Kabushiki Kaisha (SMR) Tetsudō Gijutsu Kenkyūsho. (ed.). 1943. *Tetsudō gijutsu kenkyūsho nenpō* (Annual report of research institute of railway technology), 1942 edition. Dalian: Minami Manshū Tetsudō Kabushiki Kaisha (SMR) Tetsudō Gijutsu Kenkyūsho.

Mukasa, Kingo. 1941. Kōjō no sōgokeihatsu (Promotion of interactions among factories). In *Kōjō keiei kaizen shiryō* (Materials for improvement of factory management), eds., Tokyo Kōjō Konwakai. Tokyo: Tokyo Kōjō Kanwakai.

Nihon Kōgyō Kyōkai. (ed.). 1941. *Sogo keihatsu (Mutual enlightenment)*. Tokyo: Nihon Kōgyō Kyōkai.

Nihon Kokuyū Tetsudō (JNR). (ed.). 1958. *Tetsudō gijutsu hattatsushi* (History of development of railway technology), vol. 4, rolling stock and machinery. Tokyo: Nihon Kokuyū Tetsudō.

Nihon Kokuyū Tetsudō (JNR). (ed.). 1972. *Nihon kokuyū tetsudō 100 nen shi* (A 100-year history of JNR), vol 9. Tokyo: Nihon Kokuyū Tetsudō.

Nihon Kōtsū Kyōkai. (ed.). 1972. *Tetsudō senjin roku (Records of pioneers of railways)*. Tokyo: Nihon Kōtsū Kyōkai.

Nikkan Kōgyō Shinbunsha. (ed.). 1934. *Nihon gijutsuka sōran (Directory of engineers in Japan)*. Tokyo: Nikkan Kōgyō Shinbunsha.

Railway Bureau of Governor-General of Korea. (ed.). 1934. *Chōsen tetsudō ippan* (Survey on railways in Korea). Seoul: Railway Bureau of Governor-General of Korea.

Sawai, Minoru. 2012. *Kindai Nihon no kenkyū kaihatsu taisei (National innovation system in modern Japan)*. Nagoya: The University of Nagoya Press.

Sawai, Minoru. 2015. *Teikoku Nihon no gijutsusha tachi (Engineers of imperial Japan)*. Tokyo: Yoshikawa Kōbunkan.

Taiwan Sōtokufu. (ed.). 1934. *Taiwan Sōtokufu oyobi shozoku kansho shokuinnroku (The list of staff of Governor-General of Taiwan and its offices)*. Taipei: Taiwan Sōtokufu.

Takahashi, Dankichi. 2000. *Shinkansen wo tsukkutta otoko: shima hideo monogatari (Man who manufactured bullet train: story of Hideo Shima)*. Tokyo: Shōgakukan.

Tani, Ichirō. 1982. Kenkyū kaihatsu to gakkai: kaigun kōkū tono sōgū (Research & development and academic societies: encounter naval aeronautics). In *Umiwashi no kōseki: Nihon kaigun kōkū gaishi* (Trails of naval aircrafts: history of Japanese naval aeronautics) ed., Kaikōkai. Tokyo: Hara Shobō.

Tetsudō Sharyō Gijitsu Kyōgikai. July 27, 1941. *Dai 4 kai tetsudō sharyō gijutsu kyōgikai shussekisha shimei* (Names of attendants of the fourth technical committee on rolling stock). Preserved at Diplomatic Archives of the Ministry of Foreign Affairs of Japan.

Tetsudō Gijutsu Kenkyūsho. (ed.). 1957. *50 Nenshi (50-Year History)*. Tokyo: Kenyūsha.

Tetsudōshi, Gakkai. (ed.). 2013. *Tetsudōshi jinbutsu jiten (Bibliography dictionary of railway history)*. Tokyo: Nihon Keizai Hyōronsha.

Tetsudō Daijinkanbō Jinjika (ed.). 1934 and 1941. *Tetsudōshō shokuinroku* (The list of staff of JNR). Tokyo: Tetsudōshō.

Tetsudō-shō Kōsakukyoku (The Bureau of Manufacturing of JNR). (ed.). 1925. *Dai 4 kai sharyō kenkyūkai kiroku* (The documents of the fourth conference on rolling stock). Tokyo: Tetsudōsho.

Uninscribed. 1957. Tōkaidōsen denka no omoide wo kataru zadankai (Symposium for talk of recollection of electrification of Tōkaidō line). *Denki Tetsudō* (Electric Railway). 11–4: 58–74.

Uninscribed. 1965. EF10•11•12•14 gata denki kikansha sekkei kaiko (Recollection on design of electric locomotives of EF10•11•12•14 types). *Tetsudō Pikutoriaru* (The Railway Pictorial).15–12: 16–20.

Yoshino, Shintaro. 1939. Kisha hattatsushi (Development of locomotives). *Kyōwa (Concord)* 251: 36–39.

Chapter 3
Diversification and Convergence: The Development of Locomotive Technology in Meiji Japan

Naofumi Nakamura

Abstract This chapter offers a reconsideration of the process of Japan's railway technology development during the Meiji period while focusing on the role of railway engineers and their technical imitation. It is true that underlying Japan's advancement from imitation to original design were the formation of a cadre of Japanese engineers in both the government railway and private railway companies and the manufacturing know-how cultivated through the copying of a wide variety of model locomotives. In the context of the first wave of globalization occurring around the transition from the 19th to the 20th century, the ability to freely choose from among the most advanced railway materials offered by UK, US, and Germany contributed to the diversification of technology and concomitant accumulation of experience. It was the convergence of this technology and know-how that bore fruit in the development original technology.

Keywords Railway engineers · Locomotive technology · Design skill · Japan · UK · US · Germany · First globalization

3.1 Introduction

The main purpose of this chapter is to clarify the process of railway technology development in Japan during the Meiji era (1868–1912) while focusing particularly on the role of imitation in the technology acquisition process.

One impetus for focusing on imitation is China's rapid development of high-speed railway technology in recent years. After German and Japanese high-speed railway technology was introduced in 2004, Chinese rolling-stock manufacturers accumulated know-how by meticulously imitating German and Japanese rolling-stocks and, by the 2010s, achieved levels surpassing those of

---

N. Nakamura (✉)
Institute of Social Science, The University of Tokyo, 7-3-1, Hongo, Bunkyo-ku, Tokyo 113-0033, Japan
e-mail: naofumin@iss.u-tokyo.ac.jp

M. Sawai (ed.), *The Development of Railway Technology in East Asia in Comparative Perspective*, Studies in Economic History,
DOI 10.1007/978-981-10-4904-0_3

German and Japanese manufacturers, at least in terms of speed. China has been criticized by some for "plagiarizing" or "imitating" German and Japanese technology in the development of high-speed railway technology. However, if we examine this rapid catching-up process, it is apparent that China has not simply imitated German and Japanese technology but, rather, has developed its own technology by integrating the technologies of these two countries, which differ in terms of design concept. With regard to the process by which railway systems, which comprise a wide array of technologies, are developed, it is not particularly unusual for the individual technologies that are incorporated to be meticulously copied and combined as a means of accumulating technology. 100 years ago, Japan accumulated technology in the same way. With this in mind, this paper explores the characteristics of railway technology development in East Asia through the example of Japan, which was the first East Asian country to develop original railway technology.

The process of railway technology development in Japan can be divided into four stages: (1) transfer of railway systems from the United Kingdom (UK) by the Japanese government; (2) acquisition of railway materials from the United States of America (US) and Germany and diversification of technology; (3) standardization and convergence of technology through the merger and nationalization of railway networks; and (4) development of original technology by the government railway. This paper focusses on the diversification and convergence of technology (stages 2, 3, and 4), while keeping this step-wise development in mind, and examines the process of technological diversification and convergence, primarily through the example of steam locomotives, which represent cutting-edge machine technology at the time.

Much previous research on the development of Japanese railway technology exists. For example, Yamamoto (1993) and Ericson (1996) outlined the processes by which the Japanese railway system was transferred and developed. Nakamura (1998, 2011) discussed how the foundation for railway technology was formed while focusing on engineers. However, these previous studies focused on civil engineering technology, and little analysis has been conducted on mechanical engineering technology. Therefore, in this study, I explore the formation of Japan's railway system from the standpoint of rolling-stock technology. The most important previous study on this subject is Sawai (1998). Sawai (1998) focused on the relationship between railway-owned factories and private factories during the development of the railcar manufacturing industry and pointed out the importance of the former's direct provision of technological instruction to the latter. The railway-owned factories strongly emphasized the technology development through repair work and considered the creation of new locomotives as "something to be done in between repair work." That said, insofar as the locomotive technology developed by railway-owned factories became the foundation for the shift to domestically-manufactured locomotives in the 1910s, it cannot be ignored. Therefore, in this paper, I re-examine the process by which locomotive technology was developed by government and private railways.

Another useful reference when considering the development of locomotive technology is the analysis by Nakaoka (2006) of the process by which Japan's ship building industry achieved technological independence. Nakaoka (2006) focused on the difference between basic design and detailed design (i.e. working plans) with regard to the construction of large-scale iron steamships. According to Nakaoka (2006), the goal of the Imperial College of Engineering was to nurture engineers capable of developing basic designs and of drawing general layouts; it was assumed that the ability to create working plans and subsequent skills would be acquired through on-the-job experience. In the case of mechanical engineering technology, which, in many respects, requires more technical skill than civil engineering, the ability to create detailed plans that can be used on the shop floor is the key to achieving technological independence. Thus, it will be necessary to carefully analyze the process by which such skill was developed, including the aspect of human resource development. Taking this point into consideration, in this paper, I investigate the processes by which locomotive design skills were developed by government and private railways while also paying attention to the characteristics of the mechanical engineer employed by these railways.

This paper is divided into four sections corresponding to the four stages of railway technology development, these being (1) technology transfer, (2) diversification, (3) convergence, and (4) development of original Japanese technology. In Sect. 1, I discuss the two patterns of technology transfer that were adopted; the first involving the hiring of foreign employees mainly from Britain and the second involving the hiring of consulting engineers from the US. In Sect. 2, I discuss the different processes by which government and private railways acquired technical know-how as Japan began to achieve technological independence in terms of railway technology while focusing on the diversity of technology that resulted from these divergent acquisition processes. In Sect. 3, I examine the processes by which this diverse technology was converged while paying attention to the cultivation of Japanese engineers within various structural frameworks including laws governing the operation of railways and railway nationalization policy. In so doing, I also consider the significance of imitation (reverse engineering). Finally, in Sect. 4, I clarify the process by which the first mass-production, Japanese-manufactured standard locomotive was developed while examining the selection and imitation of a model locomotive, acquisition of the ability to create working plans, and establishment of the designated manufacturer system.

3.2 Transfer of Technology

The first Japanese railway, financed and managed by the government, was opened in 1872 and ran between Tokyo (Shimbashi) and Yokohama. In this section, I examine the scheme by which railway technology was introduced into Japan during the period starting with the establishment of the first railway business (1872) up to just before the first boom in private railway companies (1886).

The decision to establish a railway business was taken by the Japanese government in December, 1869, marking the start of railway construction in Japan. Based on an introduction by Harry Smith Parkes, the Consul General of the UK to Japan, the job of laying rails and securing the necessary funds was entrusted to Horatio Nelson Lay (Hayashida 2009, 38–39). Serving as a proxy for the Japanese government, Lay was responsible for the selection of railway engineers and procurement of materials. However, the Japanese government rescinded its loan agreement with Lay in December 1870 due to a disagreement over the method by which Lay had secured funds and designated the Oriental Bank as the new proxy agent. The Oriental Bank was entrusted with supervision of the railway, while the Malcolm Brunker & Co. based in London was entrusted with procurement of materials (Hayashida 2009, 233). The foreign employees selected by Lay were retained in their original positions.

In 1871, the first ten locomotives arrived in Japan one after the other from Britain. Operation of the government railway line between Shimbashi and Yokohama was launched the following year in 1872. At the time, construction and operation of the government railway relied almost entirely on foreign employees, the vast majority of whom were British. For all intents and purposes, the director, engineer-in-chief, locomotive superintendent, secretary, and others in managerial positions had the authority to decide specifications for railway construction and for procuring materials. Furthermore, according to the agreement with the Oriental Bank, the Malcom Brunker & Co. was given exclusive rights to import materials. In this manner, a procurement route was established wherein foreign (British) employees ordered locomotives from British locomotive manufacturers through a British trading company.

However, that does not mean that the foreign employees took unfair advantage of the Japanese railway. For example, in 1882, A. S. Aldrich, who lived and worked in Japan for more than 25 years as secretary and agent of the Railway Bureau, succeeded in reducing procurements costs for the government railway by eliminating the Oriental Bank as an intermediary and directly contracting with Malcom Burnker & Co., thereby reducing handling costs from 2.5% to 1.5% (Hayashida 2009, 277–278). However, that did not change the procurement route for railway goods; even in the 1880s, Japan was still purchasing locomotives from the UK.

In contrast to the Honshu (Japan's main island) where railway construction was started with funding, personnel, and materials from the UK, railway construction in Hokkaido, which at the time was under the authority of the Hokkaido Development Commission (Kaitaku-shi), was advanced through technology transfer from the US. The individual who played a central role in this technology transfer was the American engineering advisor Joseph Ury Crawford (JNR Hokkaido so-kyoku 1976, 748–749).Crawford arrived in Japan in 1878 at the invitation of the Hokkaido Development Commission and assumed comprehensive responsibility for the construction and operation of the Horonai railway as the consulting engineer and inspector (Shavit 1990, 113). Although Crawford's tenure only lasted three years, from late 1878 to early 1881, he was able to carry out the route selection, design, materials procurement, securing of engineers and technicians, supervision of

Table 3.1 Nationalities of locomotives in Japan: 1872–1912

(Cars)

Year	British		American		Germany		Swiss		Japanese		Total	
	Number	Increase	Number	Increase	Number	Increase	Number	Increase	Number	Increase	Number	Increase
1872	10										10	
1877	36	26									36	26
1882	47	11									47	11
1887	95	48	2	2							97	50
1892	240	145	26	24	28	28					294	197
1897	484	244	282	256	55	27	3	3	11	11	835	541
1902	684	200	524	242	70	15	11	8	30	19	1,319	484
1907	966	282	908	384	160	90	11	0	95	65	2,140	821
1912	983	17	995	87	226	66	11	0	162	67	2,377	237

Source Sawai (1998) 27

construction and testing during this time. Service between Temiya and Sapporo was started in November 1880. The materials used for the railway were selected by Crawford and, for the most part, procured from the US. For this reason, it was two light locomotives manufactured by the H. K. Porter Company that arrived in Japan in 1880. These were the first US-manufactured locomotives in Japan.

Crawford did not only bring engineers necessary for railway construction from the US; he also provided opportunities for Japanese engineers trained at US universities to contribute. Specifically, Sōichirō Matsumoto and Seijirō Hirai, who were graduates of the Rensselaer Polytechnic Institute, were hired as construction supervisors for the Horonai railway. All routes were put into operation in 1883. As a result, technological independence was achieved in a shorter time in Hokkaido than in Honshu where all the employees were foreigners. Matsumoto and Hirai went on to become railway engineers capable of comprehensively overseeing railway construction and operation; both assumed directorship of Japan's government railway after the 1890s.

That said, the American technology transferred to Hokkaido did not have influence in other regions, and imports of American products remained limited until the turn of the 19th century. For this reason, as of 1887, British locomotives accounted for 99% of all locomotives in Japan and held a monopoly on the locomotive market (Table 3.1).

3.3 Technological Independence and Diversification

3.3.1 Technological Independence of the Imperial Government Railways

From the late-1870s to the early-1880s, the Imperial Government Railways (IGR) undertook rapid localization of various technologies, especially those related to civil engineering. Let us examine this point in terms of the change over time in the number of foreign engineers that were employed by the government railway (Fig. 3.1). As can be seen from the figure, the number of foreign engineers peaked at over 30 and fell dramatically in the latter half of the 1870s to under 10 in 1878.

Underlying such efforts were the cessation of foreign investments stemming from fears of being colonized and the drying up of funds for railway construction resulting from the inflationary pressures following the Satsuma Rebellion which depleted governmental coffers. Both of these created circumstances that made it difficult to hire expensive foreign employees. The replacement of foreign engineers by Japanese engineers became an urgent challenge for the survival of the government railway. Having managed to overcome this threat through the establishment of a cadre of Japanese civil engineers, the government railway took subsequent steps to further eliminate foreign dependence. In the latter half of the 1890s, British employees with the authority to procure materials, including the

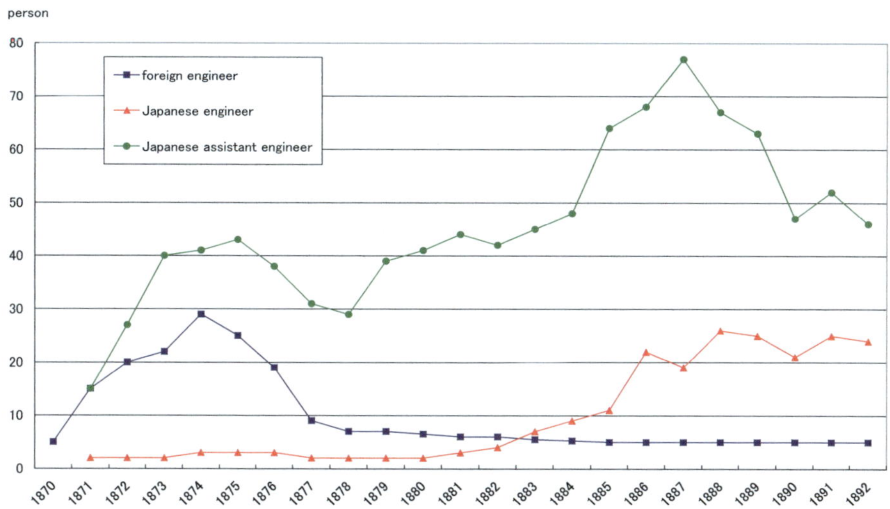

Fig. 3.1 Number of engineers in the IGR: 1870–1892. *Source* Nakamura (2016). 94

British directors, engineers in chief, and locomotive superintendents were dismissed, and the authority to select and procure railway materials was transferred from British to Japanese. Japanese engineers and managers with experience studying abroad in the US played a pivotal role in this transition.

As can be seen in Fig. 3.2, with regard to managers in the Railway Operations Bureau in the latter half of the 1890s, of the six engineers and administrative officials ranked class 3 or higher, four had experience studying abroad in the US. The eventual replacements for the hired British employees were Japanese engineers and managers who, for the most part, had studied in the US. It is also notable that the individuals responsible for materials procurement, Director Matsumoto and Tamiyoshi Zushi (head of the accounting section), both had experience studying in the US. This had a favorable impact on the inflow of American products into the railway goods market on Honshu.

Further, with regard to the mechanical engineers, we note the establishment of a locomotive department in 1896 and the appointment of Kōji Miyazaki as its chief. Miyazaki graduated from the School of Mechanical Engineering in the Imperial College of Engineering and was assigned to lead the locomotive department after serving as director of the Kobe Works. He was the first Japanese individual to head the railway section. Taking this into consideration, if we look at the change over time in composition of personnel at the government railway with some connection to locomotives (locomotive department, manufacturing plant), it is apparent that the number of Japanese mechanical engineers began to increase at the end of the 1890s (Table 3.2). As of 1900, there were 14 engineers and 64 assistant engineers in the locomotive department. Of the 14 engineers, 12 were graduates of the Imperial College of Engineering and the Imperial University (Fig. 3.2). From the foregoing

	May-94	Jul-95	Jul-96	Nov-97	Feb-99	Apr-00	Apr-01	Sep-02	education	note
S. Matsumoto	director			director					76 RPI, CE	1903 died
T.Zushi	security			head of Accounting D.					76-78, Study in USA	
K.Haraguchi	chief engineer		→						78 RPI, CE	1896 retired
S.Hirai	engineer			head of Traffic D.					78 RPI, CE	1904 director of the IGR
R.Matsuda	engineer			head of Construction D.					81 Glasrow U., CE	
M.Sengoku	engineer		→						78 U-Tokyo, CE	1896 retired
K.Miyazaki	engineer			head of Locomotive D.	→				79 ICE, ME	1894 FM of Kōbe Works, 1896 LS
M.Mori	engineer				→					1894 FM of Nagano Works
S.Hata	engineer					head of Locomotive D.			85 ICE, ME	1894 FM of Shimbashi Works, 1900 LS
K.Utsunomiya	engineer								86 ICE, ME	
G.Shiba	engineer			→			engineer		91 IU, ME	1897 Acting FM of Shimbashi Works
H.Mori	engineer								91 IU, ME	1899 FM of Kōbe Works
T.Hattori	assistant	engineer →							74 Technical Assistant	1896 retired
S.Ichikawa	assistant		engineer							
J.Yokoi				engineer					96 IU, ME	
S.Matsuno				engineer					96 IU, ME	
M.Yoshino				engineer					96 IU, ME	
I.Suzuki						engineer			90 IU, ME	1900 transferred from Hankai Railway
Y.Aoyama					assistant	engineer			97 IU, ME	
K.Nagami					assistant	engineer			97 IU, ME	
R.Kanashige					assistant	engineer			97 IU, ME	
S.Ōtsuka					assistant	engineer			97 IU, ME	
R.Nomura	assistant					engineer			91 HTS, ME	
M.Ikeda							engineer		99 IU, ME	
N.Fukushima								engineer	99 IU, ME	
Y.Shima								engineer	94 IU, ME	1901 transferred from Kansei Railway
S.Kobayashi								engineer	1900 IU, ME	

Fig. 3.2 Top managers and mechanical engineers in the Imperial Government Railways: 1894–1902. *Source* Teishin-shō (ed.) (1894–1896, 1902). Naikaku Kanpō-kyoku (ed.) (1897–1901). and Nihon Kokuyu Tetsudō (ed.) (1958). *Notes* IU = Imperial University, HTS = Tokyo Higher Technical School, FM = factory manager, LS = locomotive superintendent, → = retired or transferred from IGR

discussion, we can infer that it was the cadre of engineers who had, for the most part, graduated from mechanical engineering departments of Japanese institutes of higher education that played a leading role in achieving technological independence during the period in question.

Incidentally, what were the critical factors that enabled the locomotive department to achieve technological independence? Nakaoka (2006) explains that, when thinking about the development of mechanical engineers' ability to design locomotive, it is important to make a clear distinction between basic design and working plans. Of these two, the education provided by institutes of higher learning played an important role in nurturing engineers' ability to create basic designs, which determine the overall scheme of a railway system. In contrast, the creation of working plans requires not only theoretical training but, also, technical mastery backed by practical experience. If this distinction is indeed correct, the question of

Table 3.2 Number of employees in the IGR's locomotive department: 1894–1906

(Person)

Fiscal year	Engineer	Assistant engineer	Security	Clerk	Assistant clerk	Staff (koin)	Worker (shiyonin)	Total
1894	6	46		26	16	-	-	-
1895	7	49		26	18	-	-	-
1896	7	53		32	6	-	-	-
1897	10	51		37	5	118	5,422	5,643
1898	9	63		34	9	151	5,780	6,046
1899	12	66		36	14	141	5,934	6,203
1900	14	64	1	41	22	140	6,399	6,681
1901	15	78	1	47	3	140	6,983	7,267
1902	18	81	1	51	1	144	7,022	7,318
1903	18	81		52	1	179	7,093	7,424
1904	14	70		45		180	7,248	7,557
1905	20	97		57		306	9,671	10,151
1906	23	101		67		3,930	16,431	20,552

Source Teishin-shō (ed.) (1894–1896), Naikaku Kanpō-kyoku (ed.) (1897–1906) and Tetsudō-kyoku (ed.) (1894–1906)

how Japanese graduates of engineering schools acquired the skills needed for creating working plans becomes an important point of discussion.

One means was the creation of prototype locomotives at manufacturing plants attached to the government railway. Various types of domestically-manufactured locomotive prototypes were designed and created at the government railway Kobe Works between 1893 and 1902 under the direction of R. F. Trevithick (British, locomotive superintendent). It is through this process that Hikozō Mori (engineer, graduate of the Imperial University) and Yoshimatsu Ōta (draftsman) were nurtured as personnel capable of creating working plans. This led to the establishment in 1901 of a designer section within the locomotive department, which was filled by young mechanical engineers who were graduates of the Imperial University (Fig. 3.2). Thus, the creation of basic designs and specification documents was taken over by Japanese personnel. The government railway subsequently transferred all authority for procurement of railway materials including locomotives to Japanese staff.

3.3.2 Acquisition of Locomotive Technology by Private Railways

Next, let us examine the sudden rise in prosperity of the private railways. The first railway boom occurred in the mid-1880s, followed by a second boom in the

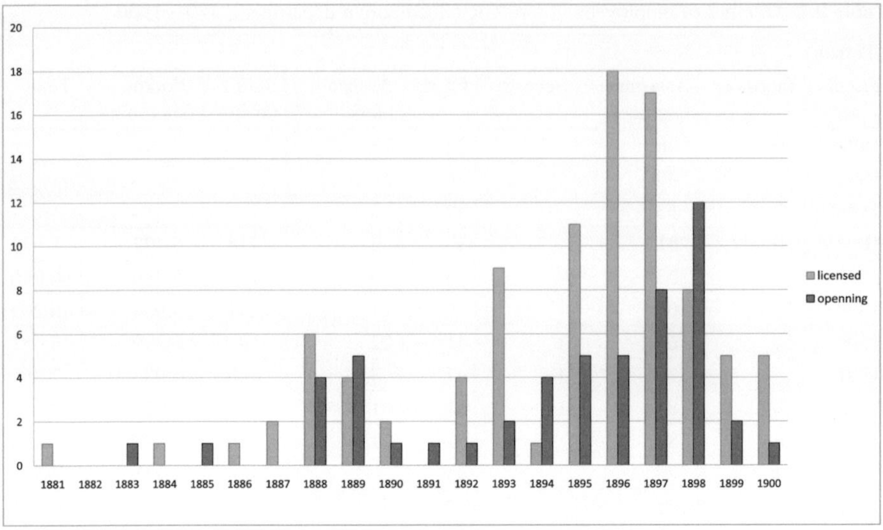

Fig. 3.3 Rising of private railways. *Source* Nakamura (2016) 92

mid-1890s, both of which led to the establishment of numerous railway companies (Fig. 3.3).

As a result, in the 1890s, the number of locomotives under private operation was greater than that under government operation (Fig. 3.4), resulting in the collapse of the government railway's monopoly on locomotive technology. With the exception of Nippon Railway, the majority of the private railway companies did not receive technical support from the government railway and had to secure engineers and railway materials on their own. Kyūshū Railway, for example, hired a German engineering consultant and procured all materials from Germany. Similarly, Sanyo Railway and other private railways in Western Japan began proactively purchasing American-manufactured locomotives at the start of the 1890s. The reason why Sanyo Railway was able to become the first railways to escape the influence of the government railway and to procure locomotives on its own had to do with the formation of a cadre of Japanese engineers centered around Hikomatsu Iwasaki (graduate of the Imperial College of Engineering). From the composition of technical personnel at Sanyo Railway as of 1900 (Table 3.3), of the 13 senior mechanical engineers, five had bachelors of engineering degrees (of which one was for electrical engineering) and two were graduates of Tokyo Higher Technical School. It should be noted that there were fewer college graduates than in the case of the government railway and that engineers such as Yaeji Nogami who rose from the ranks of draftsmen were assigned to important positions. One factor that contributed to the nurturing of large numbers of experienced master technicians at

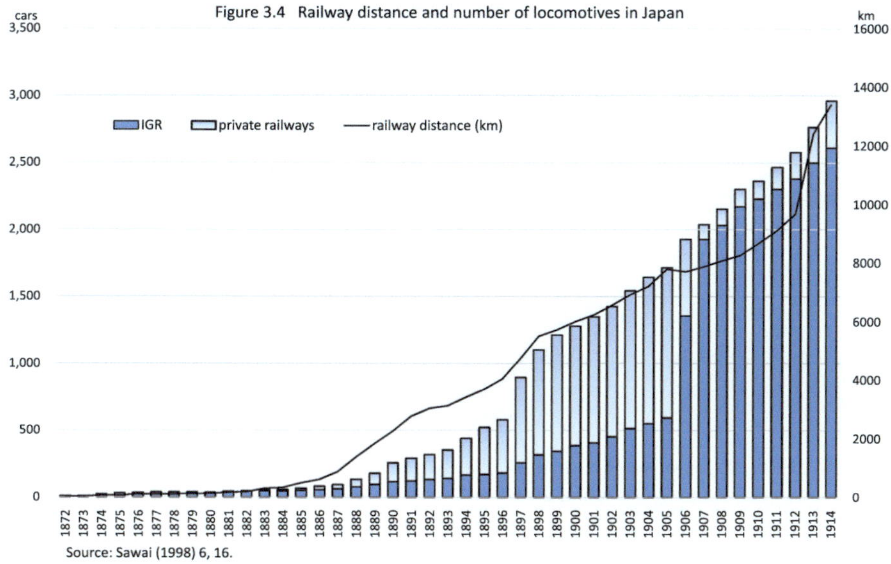

Fig. 3.4 Railway distance and number of locomotives in Japan. *Source* Sawai (1998) 6, 16

Table 3.3 Mechanical engineers in the Sanyo Railway Co. in 1900

	Section	States	Education	Note
H. Iwasaki	locomotive superintendent	Head of Loco. D.	1883 ICE, ME	1888 transferred from Hokkaido Government Railways
M. Ishihara	Driving section	Manager	1898 IU, ME	1905 chief of design section
D. Yoshino	Electric section	Manager	1896 IU, EE	
S. Tabuchi	Hyōgo works	Manager	1894 IU, ME	Chief of locomotive manufacturering
S. Inoue	Hyōgo W. woodworking shop	Chief		
S. Matsui	Hyōgo W. fitting shop	Chief	Fitter	1899 transferred from Hokkaido Coal-Mine Railway Temiya Works
E. Endo	Suma boiler shop	Chief	1891 HTS, ME	1901 manager of Takatori Works
M. Moriyama	Hyōgo W. forging shop	Chief	Founder	1897 transferred from Ōsaka Arsenal
K. Tsukamoto	Drawing office	Chief	1899 IU, ME	1901 chief of design section
T. Otsuka	Inspection section	Chief	1890 HTS, ME	
K. Arakawa	Hyōgo W. maintenance shop	Chief		
Y. Nogami	Hyōgo works	Engineer	Draftsman	1895 transferred from Kure Naval Dockyard
C. Kakimura	Hiroshima works	Manager		

Source Kinoshita (ed.) (1900). Nihon Kokuyu Tetsudō (ed.) (1958)
Notes Loco. D. = Locomotive Division, ME = mechanical engineer, EE = electric engineer
ICE = Imperial College of Engineering, IU = Imperial University, HTS = Tokyo Higher Technical School

Fig. 3.5 Tender locomotive (4-4-0) made by Sanyo Railway, Hyōgo Works (1901), (by kind permission of the Railway Museum, Japan)

Fig. 3.6 Tender locomotive (4-4-0) made by Neilson & Co (1891), (by kind permission of the Railway Museum, Japan)

Sanyo Railway was the development of numerous locomotive prototypes at its Hyogo Works. Starting with the reverse engineering of a British-manufactured locomotive in 1896, the team of engineers at Sanyo Railway centered around Iwasaki copied various types of British and America locomotives. Comparing the 2B (4-4-0) tender locomotive manufactured at the Hyogo Works in 1901 (Fig. 3.5) and the same type tender locomotive manufactured by Neilson & Co., which served as its model (Fig. 3.6), it can be seen that they are exactly the same in physical appearance. It is likely the repetition of such reverse-engineering that enabled draftsmen to acquire the skills needed for creating working plans.

Another factor contributing to the diversification in the source of locomotives was the sequential adoption of open competitive and selective bidding arrangements by both the government railway and private railways, which led to the creation of a competitive materials procurement system. For example, the Sanyo

Railway's "goods procurement regulations" allowed for both selective bidding and negotiated (limited-tender) contracts, indicating that competitive bidding was also being used by private companies. The intensification of competition in terms of price and lead time brought about by the introduction of competitive bidding was favorable to American and German manufactures that had succeeded in reducing costs and manufacturing times through the development and implementation of the American-style interchangeable parts manufacturing. In contrast, the British, who were fixated on high quality production through skilled manual work, were unable to compete in terms of cost and lead time. As a result, in 1897, the number of new locomotives purchased from the US exceeded that from the UK; the share of new locomotives purchased from Germany also began to increase in the 1900s (Table 3.1).

3.3.3 Diversification of Locomotives

In addition, intensification of international competition in supply of railway materials resulting from the first phase of globalization enabled both the government railway and private railways to optimize procurement of locomotives on a global scale. Specifically, this led to a dramatic increase in the purchase of American-made locomotives by private railways (Table 3.1). Concerns regarding the circumstances at the time and the penetration of American-made locomotives into the Japanese market are expressed in the following report by the British Consular to Japan dated July 16, 1894:

> For railway locomotives, of which there has also been an increased import, the Japanese private railway companies appear inclined to have recourse rather to the United States than to England; and in one recent instance at least a private company disposed of several of their English built engines to the Government railways, supplying their place with newly-imported American engines at a cost greatly in excess of the proceeds realised by the sale of the old ones (Foreign Office 1894, 9).

In the midst of the second railway boom, the British Secretary of Legation in Tokyo, Gerard Lowther, composed two detailed reports titled *Report on the Railways of Japan* in 1896 and 1897, which he sent back to his home country. In these reports, Lowther describes the circumstances surrounding the railway boom in Japan and, in addition to warning British locomotive manufacturers not to miss this business opportunity (Lowther 1895), and depicts the Japanese market in the context of the railway boom as follows:

> In my Report No. 390, of 1896, I pointed out the fact that a very large proportion of the locomotives on Japanese railways were of English make. I believe the proportion to have been at a comparatively recent date, English 224; American 6. Some of the latter were purchased in 1890, but did not give satisfaction, the finish being unsatisfactory and the consumption of coal greater than on the English engines. However, some more were ordered in 1894, with a like result. With this experience, it is not to be wondered at that the

Japanese should prefer to purchase English made locomotives, but we do not have a monopoly in this line (Lowther 1897).

Here, we see Lowther emphasizing the dominance of British-made locomotives in the Japanese market as well as the comparative advantage over American-made locomotives in terms of fuel consumption and, in general, expressing an optimistic outlook. At the same time, however, he points out that the British monopoly in the Japanese market is starting to crumble and urges the British manufacturers to prepare for coming competition in the locomotive export market.

The US was not the only challenger to British dominance. Among the railways that were newly established during the first railway boom, some companies such as the Iyo Railway and the Kyūshū Railway imported technology from Germany. For example, at the same time as its establishment in 1888, Kyūshū Railway hired a German engineering consultant, Hermann Rumschöttel, and purchased the full array of railway materials including rails, bridge materials, and locomotives from the Gutehoffnungshütte Ironworks via the C. Illies & Co. trading company. In turn, the Gutehoffnungshütte Ironworks outsourced the manufacture of locomotives to the Hohenzollern and Krauss companies (Usui 1973, 211).

With respect to the importation of German technology by Kyūshū Railway, Yuzawa (1991) emphasizes the importance of political factors underpinned by diplomatic pressure from the German government (Yuzawa 1991, 207–210). However, comparing the price of tank locomotives manufactured by the Krauss Company to British-made tank locomotives in the same class (weight) as of 1888–1889 (Table 3.4), it can be seen that both of the German-made locomotives were slightly less expensive than the British-made locomotives. That is to say that the German-made locomotives already held a competitive advantage over British-made locomotives in terms of price at this point. Thus, the political pressure from the German government at the point when the Kyūshū Railway was deciding which locomotive to purchase likely amounted to nothing more than a demand to include German products in the list of products being evaluated. Indeed, it appears that the

Table 3.4 Comparison between British and Germany tank engines

	Year	1888	1889
Krauss (Germany)	Axle arrangement	0-4-0	0-6-0
	JNR type No.	10 type	1440 type
	Price (£)	920	1,315
	User name	Kyūshū Railway	Kyūshū Railway
	Weight (t)	25.5	34.61
Nasmyth Wilson (British)	Axle arrangement	0-6-0	2-4-2
	JNR type No.	1100 type	600 type
	Price (£)	1,030	1,336
	User name	Sanyo Railway	IGR
	Weight (t)	21.34	31.32

Source Nasmyth'Loco. Specifications', 'Reference Book 2', and Krauss 'Auftragsbuch'

Kyūshū Railway was satisfied with the price and quality of the German locomotives. In fact, Kyūshū Railway continued to purchase locomotives made by the Krauss Company even after engineering consultant Rumschöttel stepped down in 1892 (Tetsudo-shi Gakkai 2013, 451–452), all the way up to 1898. As a result, as of 1897, 98% of the German-made locomotives in Japan (excluding light railways) were made by the Krauss Company (Nakamura 2016, 89).

If we consider the diversification of locomotives, whereas, in 1887, all but two locomotives in Hokkaido were British-made (the other two being American-made), by 1897, the number of American-made locomotives had increased to 33.8%, and the number of manufacturing countries had diversified to include Germany, Switzerland, and Japan (Table 3.1). As a result of the government railway and private railways independently procuring railway materials from various countries around the world, a great variety of different locomotives with varying specifications existed in Japan around the turn of the 19th century,

3.4 Convergence of Technology

3.4.1 Starting Points for Standardization

Around the turn of the century, efforts to organize institutes of higher education and vocational schools to cultivate engineers were advanced, and a full-fledged effort was begun to expand the ranks of Japanese engineers. Factories affiliated with railways (e.g. IGR's Kobe Works, the Sanyo Railway's Hyogo Works, and the Nippon Railway's Omiya Works) and independent rolling-stock manufacturers (e.g. Kisha-seizo Co.) began producing copies of the foreign-made locomotives. Each factory accumulated technology in the process of performing maintenance on the locomotives, and increased their manufacturing skill by meticulously copying existing locomotives.

At the same time, efforts were being made to converge technology that had diversified. In 1900, the list of railways in Japan comprised the government railway and 43 private railways. Just to traverse Honshū it was necessary transfer between the Nippon Railway, the Tokaidō Line of the Imperial Government Railways, and the Sanyo Railway. As such, for long-distance transportation, joint through traffic involving several different railways was unavoidable, and the relative importance of such transportation increased with the development of the railway transportation. From this standpoint (Table 3.5), joint through traffic of both passengers and freight started in earnest in 1895, after the Sino-Japanese War. In particular, after 1899, joint through traffic of freight involving both the government railway and private railways accounted for more than 20% of total transportation volume.

With the full-fledged expansion of joint through traffic following the end of the Sino-Japanese war, it became necessary to regulate the capacity and rail weight of individual railways as well as the strength of couplers. The Railway Operation Act,

Table 3.5 Joint through traffic related to the IGR: 1893–1903

	Fiscal year	1893	1894	1895	1896	1897	1898	1899	1900	1901	1902	1903
Passenger (1000 person)	Total volume, arrive (A)	14,439	14,736	18,090	22,681	28,098	31,552	28,208	31,339	31,199	31,067	33,271
	From PR to IGR (B)	386	378	604	1,352	1,298	1,615	1,623	1,763	1,596	1,737	2,044
	JTT retio (B/A)	2.7%	2.6%	3.3%	6.0%	4.6%	5.1%	5.8%	5.6%	5.1%	5.6%	6.1%
	Total volume, departure (C)	14,423	14,775	18,282	22,183	27,804	31,314	28,156	31,460	31,596	31,402	33,478
	From IGR to PR (D)	370	417	796	854	1,004	1,377	1,571	1,884	1,993	2,073	2,252
	JTT retio (D/C)	2.6%	2.8%	4.4%	3.9%	3.6%	4.4%	5.6%	6.0%	6.3%	6.6%	6.7%
Fright (1000 ton)	Total volume, arrive (E)	958	908	1,134	1,277	1,381	1,569	1,872	2,032	1,895	2,232	2,454
	From PR to IGR (F)	8.9	16	24	33	40	48	52	50	37	82	99
	JTT retio (F/E)	0.9%	1.8%	2.1%	2.6%	2.9%	3.1%	2.8%	2.5%	2.0%	3.7%	4.0%
	Total volume, departure (G)	1,077	1,010	1,304	1,446	1,555	1,805	2,306	2,642	2,516	2,972	3,245
	From IGR to PR (H)	128	118	194	202	214	284	486	660	658	822	890
	JTT retio (H/G)	11.9%	11.7%	14.9%	14.0%	13.7%	15.7%	21.1%	25.0%	26.1%	27.6%	27.4%

Source Naikaku Tōkei-kyoku (ed.) (1893–1896). Tetsudō Sagyō-kyoku (ed.) (1897–1903)
Note The total volume of departure (C and G) in 1895 and 1896 was estimated by A−B + D and E−F + H

which was promulgated in 1900, contained ancillary regulations related to construction, operations, signals, traffic, and statistics. For example, the construction regulations contained specific rules regarding railway gauge, slope, curve radius, rail weight, bridges and tunnels, construction and rolling-stock limits, and the shape of couplers in order to eliminate obstacles to smooth linkage between the government and private railways. Compliance with these regulations enabled the direct linking of government and private railways. That said, with respect to rolling-stocks, the construction regulations only went so far as to set rolling-stock limits but did not attempt to unify formats or to standardize specifications. However, the joint through traffic at the time was accomplished by the sharing of rolling stock between the government railway and private railways in the form of "temporary loans of rolling stock"(Tetsudō-in Kensetsu-bu 1909, 32). As such, it was necessary to establish certain standards not only to unify coupler shape and track load capacity but, also, in terms of the design of the rolling stock itself.

In 1901, an office was established in the design section of the Railway Bureau to formulate standards for rolling-stock that could also be applied to private railways. To deal with "issues related to the design and standards for rolling-stock and railway equipment," the office assembled a team of engineers that included not only engineers from the government railway such as Seikichirō Hata (locomotive superintendent) and Gontarō Shiba (director of the Shimbashi Works) but also engineers with substantial experience in private railways such as Ikuyata Suzuki (former chief of the locomotive section of the Hankai Railway) and Yasujirō Shima (former chief of the locomotive section of the Kansei Railway) (Fig. 3.2). Under the direction of the government, the team attempted to unify standards for locomotives.

Meanwhile, mergers between private railways began to increase in frequency around the turn of the century, and standardization of railcars was advanced in each region, with trunk railways such as the Nippon Railway, Kansei Railway, Sanyo Railway, and Kyūshū Railway playing a leading role. For example, the previously mentioned Kyūshū Railway, which started with the importation of German technology, merged with the Chikuhō Railway in 1897 and began purchasing locomotives from the US. Chikuhō Railway, which transported coal had been using American-made locomotives with excellent pulling power. The reborn Kyūshū Railway began to establish locomotive standards based on these American-made locomotives. Similarly, in addition to merging with nearby branch railway companies, trunk railways such as the Nippon Railway, Kansei Railway, and Sanyo Railway advanced standardization of their own railcars.

3.4.2 *Nationalization and Convergence of Locomotive Technology*

The advancement of standardization experienced a sudden boost with the nationalization of railways in 1906–1907, which led to the full-fledged convergence of railway technology. The Agency of Imperial Railways (which became the Agency of Railways), which was formed through the merger of 17 major private railways and the government railway, became a monopoly, owning 90% of the railway lines and 95% of the locomotives in Japan. However, the newly inaugurated the Agency of Railways was saddled with the great variety of locomotives that had been inherited from both the government railway and private railways and was far from being able to carry out unified railway operations. Accordingly, as a first step, the Agency of Railways developed a unified format for rolling-stocks and attempted to collect information regarding the overall situation. Based thereupon, the Agency of Railways established a manufacturing division tasked with the expert design and supervision of rolling-stock within the transportation section and began to prepare for the standardization of rolling-stock.

With the nationalization of railways, Japan's main railway and rolling-stock engineers began to assemble at the Agency of Railways. Table 3.6 shows the composition of engineers at Agency of Railways headquarters and in the manufacturing sections and main factories (Ōmiya, Shimbashi, Kobe-Takatori) of the respective administrative bureaus of each railway in 1909. It can be seen from the table that major railways retained substantial influence in their respective regions, with Eastern region = Nippon Railway, Central region = the government railway, Western region = Sanyo railway, Kyūshū = Kyūshū Railway, and Hokkaido = Hokkaido Coal-mine Railway. Furthermore, in the Eastern and Central regions, the plant directors also served as directors of manufacturing sections in the respective administrative bureaus and are believed, in essence, to have inherited the work of engineers attached to the plants. In contrast, the manufacturing section of the transportation division was headed by Yasujirō Shima, who had no experience in the government railway, and was also organizationally distant from the manufacturing plants attached to railway companies. With the exception of Yaeji Nogami who had worked his way up from the rank of draftsman, seven of the eight engineers in the manufacturing section were graduates of the Imperial University. They came from a variety of railways including the government railway (2), Nippon Railway (2), Sanyo Railway (1), and new graduates (2). In this manner, Manufacturing Section Chief Shima assembled university-educated engineers with proven track records, regardless of their initial railways, with the goal of improving the ability of the Agency of Railways to design locomotives.

3.5 Development of Original Japanese Technology

3.5.1 Seeking Models for Domestic Production

The impetus for manufacturing locomotives domestically came from the Agency of Railways's policy for purchasing domestically-manufactured rolling-stocks, which was adopted in 1909. The policy was the result of strong appeals from domestic

Table 3.6 Mechanical engineers in manufacturering sections of the Railway Agency in 1909

Name	Section	Status	Education	Position at the nationalization
Y. Shima	MS of Traffic department	Head	94 IU, ME	Engineer of design section, Railway Bureau
Y. Aoyama	MS of Traffic D. (08 study abroad)	Engineer	97 IU, ME	Chief of design section, IGR
M. Ikeda	MS of Traffic D. (09 study abroad)	Engineer	99 IU, ME	Engineer of driving section, IGR
S. Takasu	MS of Traffic D.	Engineer	97 IU, ME	Chief design engineer, Nippon railway Ōmiya works
S. Akiyama	MS of Traffic D. (09 study abroad)	Engineer	02 IU, ME	Chief inspect engineer, Nippon railway Ōmiya works
Y. Nogami	MS of Traffic D.	Engineer	Draftsman	Acting chief of design section, Sanyo Railway
T. Eba	MS of Traffic D.	Engineer	04 IU, ME	Assistant engineer, IGR
H. Kohaze	MS of Traffic D.	Engineer	04 IU, ME	
G. Shiba	MS of east division, manager of Ōmiya Works	Head	91 IU, ME	Manager of Nippon railway Ōmiya works
T. Okabe	Ōmiya works	Engineer	87 HTS, ME	Chief product engineer, Nippon railway Omiya works
M. Isoya	Ōmiya works	Engineer	93 HTS, ME	Assistant engineer of Nippon railway Ōmiya works
H. Mori	MS of central division, manager of Shimbashi Works	Head	91 IU, ME	Manager and chief design engineer of IGR Shimbashi works
T. Shimada	MS of central division	Engineer	00 IU, ME	Engineer, Nippon railway Ōmiya works
K. Nagami	Shimbashi works	Engineer	97 IU, ME	Engineer, IGR Nagano works
N. Fukushima	Shimbashi works	Engineer	99 IU, ME	Engineer of driving section, IGR
S. Tabuch	MS of west division	Head	94 IU, ME	Chief of design section, Sanyo railway and manager of Hyōgo works
H. Araki	MS of west division	Engineer	05 IU, ME	
E. Endō	Takatori works	Engineer	91 HTS, ME	Manager of Sanyo railway Takatori works
M. Moriyama	Takatori works	Engineer	Founder	Chief of forging shop, Sanyo railway Hyōgo works
S. Matsui	Kobe works	Engineer	Fitter	Chief of fitting shop, Sanyo railway Hyōgo works
S. Nakata	Kobe works	Engineer	04 KIU, ME	

(continued)

Table 3.6 (continued)

Name	Section	Status	Education	Position at the nationalization
M. Kanō	MS & DS of Kyūshū division	Head		Chief of locomotive manufacturing, Kyūshū railway
M. Miyazaki	MS of Kyūshū division	Chief	99 IU, ME	Chief of design section, Kyūshū railway
S. Watanabe	MS of Hokkaido division	Head	99 IU, ME	Manager of Hokkaido coal mine railway Iwamizawa works
M. Shigemi	MS of Hokkaido division	Chief	00 IU, ME	Engineer of rolling-stock section, Hokkaido coal mine railway

Source Tetsudō-in Kensetsu-bu ed. (1909). Kinoshita (ed.) (1906). and Naikaku Kanpō-kyoku (ed.) (1906, 1909)

Notes MS = manufactureing section, ME = mechanical engineer, ICE = Imperial College of Engineering

IU = Imperial University, HTS = Tokyo Higher Technical School, KIU = Kyoto Imperial University

rolling-stock makers who saw the monopsony (monopoly of demand) due to nationalization of the railway as a prime opportunity. At the same time, the government, which was plagued with a severe deficit in the international balance of payments due to a large scale issuance of foreign-currency bonds for the Russo-Japanese war, was being pressed to avoid, as much as possible, purchasing materials from overseas. In this context, the Agency of Railways revised its policy of emphasizing imported locomotives in favor of cultivating domestic makers and put forth a policy of manufacturing locomotives domestically.

Taking advantage of this opportunity, the manufacturing section of the Agency of Railways led by Yasujirō Shima took on the task of developing a large-scale, high-speed locomotive intended for trunk line railways. Specifically, the section imported the most advanced locomotives from the US, Germany, and the UK and attempted to absorb manufacturing technologies by meticulously studying and coping their design.

In 1910, competitive bidding was carried out for a superheated 2C (4-6-0) tender passenger locomotive and a BB (0-4-4-0) mallet-type compound locomotive for steep grades, which were to be used as models for domestically manufactured locomotives. Six manufacturers, including American (American Locomotive Co., Baldwin), German (Borsig, Schwarzkopff), and British (North British Locomotive, Beyer Peacock) were designated. The two German makers submitted designs that were exactly to specifications, the ALCO (American) submitted a design that replaced the axle arrangement with that of the 2C1 (4-6-2), and the NBL (British) submitted a design for a saturated steam engine rather than a superheated steam engine. In the end, separate contracts were signed with each maker.

Through this series of orders, the Agency of Railways was seeking to obtain not only the model locomotives themselves but, also, information regarding the manufacturing process. To this end, the chief of the manufacturing section, Yasujirō Shima, who had played a central role in the ordering of model locomotives, and junior engineer Kiichi Asakura were dispatched as inspectors to Berlin, where the two German manufacturers, Borsig and Schwarzkopff, were located with the goal of obtaining information regarding the "practice" of locomotive manufacturing. As soon as the order was placed, Shima and Asakura immediately began visiting the Borsig and Schwarzkopff factories on a daily basis and checked every drawing for the locomotive design. As can be seen from the statement below, what the two emphasized were the working plans for the manufacture of each part used at the factory.

> Although work is proceeding at a fast pace at each plant, each working plan must be approved by inspector Shima. As such, although we had to go to the plant as often as possible, because the two factories were quite far apart, at best, we were only able to visit each factory every other day. A record of the discussions held at each factory were delivered to Mr. Shima the next morning…Looking at the drawings that were submitted for approval, in both companies, the order in which the drawings were submitted appears to be random. Although this is inconvenient for anyone wishing to investigate the design, the drawings for parts requiring the longest time to manufacture must be delivered to the factory floor first. With respect to assembly drawings, we rely on the experience of specialized factories. Leaving this to later, although we approved working plans, these require a high level of mastery as a designer and represent highly skilled work (Asakura 1958, 10–11).

The fact that they were able to visit two of the leading rolling-stock manufacturing plants on a daily basis and to observe the "practice" in detail was a very important experience not only in terms of preparing for the domestic production of a superheated steam engine but, also, in terms of learning about factory management. On this point, the two engineers' use of their position as the placer of the orders to absorb manufacturing know-how is, in many respects, similar to the way in which shipbuilding skills were developed by the navy and the Mitsubishi's Nagasaki shipyards (Nakaoka 2006). The working plans that Shima and Asakura obtained in Germany went on to play an important role in the transition to domestic manufacture of locomotives.

3.5.2 Toward Domestic Manufacture of Locomotives

Locomotives ordered from the US and Germany began arriving in Yokohama one after the other between June and July 1911. The Agency of Railways immediately began assembling the locomotives. At the same time, engineers from the Agency of Railways, along with engineers from the Kawasaki Shipyard and the Kisha-seizō Co. (steam locomotive manufacturing company), conducted material hardness tests to evaluate working accuracy. In addition, exact records were made of each part, and efforts were made to get ready to manufacture an exact copy of the locomotive. In preparation for manufacturing a superheated steam locomotive, the Kawasaki Shipyard expanded its facilities, increased its workforce, and obtained a patent for a Schmidt-type superheated steam engine.

From 1911 to 1912, the Agency of Railways, working with private rolling-stock manufacturers, set about reverse engineering and meticulously copying the model locomotives at the parts level, and succeeded in creating a copy of a locomotive for large passenger trains. Thereafter, in 1914, it succeeded in developing a locomotive conforming to Japanese standards (Japan National Railway number 8620) that was modeled after a German-made locomotive but incorporated original design elements, and efforts to mass produce this locomotive were begun. Thereafter, various locomotives conforming to Japanese standards based on original designs were successively developed, and Japan entered the age of locomotive manufacturing based on original technology.

3.6 Conclusion

Finally, let us summarize the process by which railway technology was developed in Japan in terms of the four stages of technological transfer, these being (1) technology transfer, (2) diversification, (3) convergence, and (4) development of original Japanese technology.

First, let us look at the Stage (1). From the closing days of the Tokugawa shogunate up to the early Meiji era, railway technology was transferred to Japan by two means: (1) through the hiring of foreigners (Oyatoi) or (2) through general contracts. The former method was promoted by the British for railways on Japan's main island (Honshū). The latter method was encouraged by the US and employed by the Hokkaido Development Commission for railways in Hokkaido. Of these two, the hiring of foreigners had greater influence on the development of Japanese railway technology. Therefore, previous studies have focused on this method. However, the second method should not be overlooked insofar as it was responsible for the nurturing of Soichirō Matsumoto and Seijirō Hirai, engineers who played a central role in the Imperial Government Railways during from 1890s to 1900s, and for establishing the foundation for subsequent diversification. However, in the early Meiji era, the influence of the hired foreign employees in Honshū and the British who had a monopoly on the supply of railway materials was extremely strong; the share of American locomotives was on the order of 1%.

In stage (2), the source countries of locomotives in Japan diversified at a rapid pace. The rise in prosperity of railway companies and technological independence by Japanese engineers that occurred in the 1880s and 1890s led to the diversification of railway technology in Japan. This diversification occurred against a backdrop of intensifying international competition in the global market for locomotives around the turn of the century and the possibility of optimally purchasing railway materials from around the world that emerged as a result of the first stage of globalization.

Within Japan, competitive bidding and limited competitive bidding were introduced sequentially starting in the 1890s, leading to the creation of a competitive system for procuring materials. At the same time, because each railway company purchased locomotives independently, there were many locomotives in Japan with a great variety of specifications. Furthermore, as part of this process, efforts to manufacture copies using various locomotives from different countries as models began, leading to the accumulation of rolling-stock manufacturing technology.

In stage (3), the lack of uniform standards among the different railways emerged as a problem as joint through traffic began to increase, resulting in the need for standardization of railway facilities. Following the promulgation of the Railway Operations Law in 1900, various regulations were adopted, starting with construction regulations, aimed at standardizing the rails, couplers, and the rolling-stock itself. After 1900, the merger of private railways began to increase, leading to standardization of rolling-stock in each region. This move towards standardization was propelled forward with the nationalization of railways in 1906–1907, resulting in full-fledged convergence of railway technology.

In stage (4), a manufacturing section was established within the Agency of Railways, and efforts were begun together with private rolling-stock manufacturers to make locomotives in Japan. As part of that effort, various locomotives to be used as models for domestic manufacture were ordered from Europe and the US, and manufacturing know how was acquired by taking advantage of the inspection system. As a result, by 1914, Japan had developed the ability to independently design and manufacture locomotives, which ultimately led to the emergence of mass-produced Japanese-made locomotives.

It is true that underlying Japan's advancement from imitation to original design over such a short period of time were the formation of a cadre of Japanese engineers in both the government railway and private railway companies and the manufacturing know-how cultivated through the copying of a wide variety of model locomotives. In the context of the first wave of globalization occurring around the transition from the 19th to the 20th century, the ability to freely choose from among the most advanced railway materials offered by different countries around the world contributed to the diversification of technology and concomitant accumulation of experience. It was the convergence of this technology and know-how that bore fruit in the development of original technology.

Acknowledgements This research was supported by funding from the research grant of Japan Center for Economic Research.

Primary Sources

Krauss-Maffei AG. Krauss auftragsbuch. Bayerisches Wirtschaftsarchiv.
Nasmyth Papers. Locomotive specifications. Salford City Archives.
Nasmyth Papers. Reference book 2. Salford City Archives.

References

Asakura, Kiichi. 1958. *Gijutsu seikatsu 50-nen (My 50-years career as engineer)*. Tokyo: Nikkan Kōgyō Shinbun-sha.

Ericson, Steven J. 1996. *The sound of the whistle: Railroads and the state in Meiji Japan*. Cambridge Mass.: Harvard University Press.

Foreign Office (UK). 1894. Report on the foreign trade of Japan for the year 1893. In *Diplomatic and Consular Reports. 16 July 1894*. London: Her Majesty's Stationary Office.

Hayashida, Haruo. 2009. *Nihon no tetsudō sōsō-ki (In the pioneer days of Japan's railways)*. Kyoto: Minerva Sho-bō.

JNR Hokkaido-sokyoku ed. 1976. *Hokkaido tetsudō 100 nen-shi jyō* (100-year history of Hokkaido railways, vol. 1). Sapporo: Japan National Railway, Hokkaido Division.

Kinoshita, Ritsuan (ed.). 1900. *Teikoku tetsudō yōoukan Dai-1-pan (Handbook of railways in Imperial Japan ver. 1)*. Tokyo: Tetsudō Jihō-kyoku.

Kinoshita, Ritsuan (ed.). 1906. *Teikoku tetsudō yōkan dai-3-han (Handbook of railways in Imperial Japan ver. 3)*. Tokyo: Tetsudō Jihō-kyoku.

Lowther, Gerard. 1895. Report on the railways of Japan. In *Foreign Office 1896 miscellaneous series* (Consular reports on subjects of general and commercial interest). No. 390. London: Her Majesty's Stationary Office.

Lowther, Gerard. 1897. Report on the railway of Japan. In *Foreign Office 1897 miscellaneous series* (Consular reports on subjects of general and commercial interest). No. 427. London: Her Majesty's Stationary Office.

Naikaku Kanpō-kyoku ed. 1897–1906. *Shokuin-roku* (A directory of government officials). Tokyo: Naikaku Kanpō-kyoku.

Naikaku Tōkei-kyoku ed.. 1893–1896. *Nihon teikoku tōkei nenkan* (Annual statistic of the Imperial Japan). Tokyo: Tokyo Tōkei Kyo-kai.

Nakamura, Naofumi. 1998. *Nihon Tetsudō-gyō no keisei (Formation of Japan's railways: 1869–1894)*. Tokyo: Nihon Keizai Hyōron-sha.

Nakamura, Naofumi. 2011. Railway engineer's group in the early Meiji Japan. *Japanese Research in Business History* 28: 11–33.

Nakamura, Naofumi. 2016. *Umi wo wataru kikansha (Trading locomotives: The first globalization and the development of Japan's railways, 1869–1914)*. Tokyo: Yoshikawa Kōbun-kan.

Nakaoka, Tetsurō. 2006. *Nihon kindai gijyutsu no keisei (Formation of modern technology in Japan)*. Tokyo: Asahi Shinbun-sha.

Nihon Kokuyu Tetsudō (Japan National Railways) ed. 1958. *Tetsudō gijyutsu hattatsu-shi VI sharyō to kikai (1) (2)* (A history of development of railway technology in Japan 4 rolling-stock and machinery(1)(2)). Tokyo: Nihon Kokuyu Tetsudō.

Sawai, Minoru. 1998. *Nihon tetsudō sharyō kōgyō-shi (A history of Japanese railcar manufacturing industry)*. Tokyo: Nihon Keizai Hyōron-sha.

Shavit, David. 1990. *The United States in Asia: A historical dictionary*. Westport: Greenwood Press.

Teishin-shō (Ministry of Communications and Transportation) ed. 1894–1896. *Teishin-shō Shokuin-roku* (A directory of government officials in the Ministry of Communications and Transportation). Tokyo: Teishin-shō.

Tetsudō-in Kensetsu-bu (Construction Division, Agency of Railways) ed. 1909. *Tetsudō hōki ruijyu* (Status book of railways in Japan). Tokyo: Agency of Railways.

Tetsudō-kyoku ed. 1894–1906. *Tetsudō-kyoku nenpō* (Annual reports of Railway Bureau). Tokyo: Tetsudō-kyoku.

Tetsudō Sagyō-kyoku ed. 1897–1903. *Tetsudō sagyō-kyoku nenpō* (Annual report of Railway Operation Bureau). Tokyo: Tetsudō Sagyō-kyoku.

Tetsudō-shi Gakkai (Railway history association Japan) ed. 2013. *Tetsudō-shi jinbutsu jiten* (Who's who in Japanese railway history). Tokyo: Nihon Keizai Hyōron-sha.

Usui, Shigenobu. 1973. *Kikansha no keifu-zu 2* (Genealogy of locomotives vol. 2). Tokyo: Kōyu-sha.

Yamamoto, Hirofumi (ed.). 1993. *Technological innovation and development of transportation in Japan*. Tokyo: United Nations University Press.

Yuzawa, Takeshi. 1991. The transfer of railway technologies from Britain to Japan, with special reference to the locomotive manufacture. In *International technology transfer, Europe, Japan and the USA, 1700–1914*, ed. David Jeremy. Aldershot, Hants: Edward Elgar. 199–218.

Chapter 4
Railway Technology of South Manchuria Railway and Workers in China

Chaisung Lim

Abstract As a result of the Russo-Japanese War, railroad construction was undertaken to set up the South Manchuria Railway (SMR) network in the north-eastern part of China. However, the SMR had no option of buying railroad cars from abroad, because this was Japan's first experience with foreign railroad management of a normal-gauge railway, and the machine industry in Japan had not grown sufficiently to support it. Even though railway technology was introduced from the Western world, the SMR began to pursue its own technology and was able to create a stable profit structure. The SMR's accumulation of technology became world-class in the prewar period, and this level of development was also present in the organization's internal management. In addition, the most advanced technology was developed, due to the organization's superior management. Wartime technology was developed in response to the lack of resources experienced during the war, but it was not able to solve new technical problems, such as dieselization. The core technology was confined to Japanese employees and technology spillovers to Chinese employees were restricted.

Keywords South manchuria railway · Railway technology · Technology transfer · Railroad factory

4.1 Introduction

The purpose of this study is to examine the railway business of Mainland China, established by Japanese initiative after the Russo-Japanese War, and to discover how technology transfer, its indigenization, and further technological progress were realized, focusing mainly on track facilities and railroad cars.

C. Lim (✉)
Faculty of College of Economics, Rikkyo University, 3-34-1 Nishi-Ikebukuro, Toshima-ku, Tokyo 171-8501, Japan
e-mail: wajinsup@hotmail.com

M. Sawai (ed.), *The Development of Railway Technology in East Asia in Comparative Perspective*, Studies in Economic History,
DOI 10.1007/978-981-10-4904-0_4

The end of the Russo-Japanese War caused the decline of Russian influence and enabled the Japanese colonization of Korea. The southern lines of the Chinese Eastern Railway of the Russian Empire began to function as a railway network for the Japanese Empire, with the Gyeongbu and Gyeongui Lines connecting to the Korean Peninsula. The network of the South Manchuria Railway (SMR) was set up in Northeast China, and this railroad construction greatly contributed to the maintenance of the imperial order. However, railroad materials and cars for the SMR were bought from abroad because this was Japan's first experience with foreign railroad management of international-standard-gauge railways, and the Japanese could not access railway metal and machinery within their country. Railway technology was imported from the Western world, and both the Japanese National Railways (JNR) and private railroads were managed. However, Japan's foreign railway business began running before the domestic production of track facilities and railroad cars was accomplished.

As technology transfer progressed, the SMR began to produce its own technology. In a short time, the SMR could attract many passengers and significant freight traffic, achieving high overall productivity. The SMR, including the Fushun coal mine business, was also able to realize a sounder profit structure than the JNR, which had stable management immediately after its establishment, as well as the Korean National Railways (KNR), which was troubled by management instability until the 1930s when colonial industrialization spread on a large scale (Lim 2013). The SMR took advantage of technology transfer from the Western world, reaching early stability, and then tried to produce its own railroad cars. The SMR also participated in the world competition for the limited express passenger train with a project for the "Asia Express" services in the 1930s. The service of the limited express passenger train reached North Manchuria via Korea from inland Japan, with the establishment of Manchukuo, and reached Beijing after the outbreak of the Second Sino-Japanese War.

What kind of technology transfer supported the successful business development of the SMR? Compared to the European and American railroads, what was the level of technology achieved by the SMR? What Japanese-style features characterize it? Local employees had to acquire new railroad technology for post-war railroad management. What was the main difficulty they faced? Based on these critical issues, this study focuses on railroad construction and new cars provision as an index of railway technology and examines the evolution of the railway business in the Northeastern part of China.

The existing research related to these questions mainly consists of manuals focusing on railway vehicle technology (Ichihara et al. 1971; Ichihara et al. 1972; Ichihara 2010; Amano 2012). A notable exception is Yasutaka Takahashi (1995), who showed that, even though the SMR had no choice but to purchase many vehicles from the US upon establishment of its railway business, it soon started to produce its own locomotives by copying the "American type" locomotives. The SMR secured a supply of locomotives from railway workshops or Japanese domestic manufacturers beginning in the 1920s, but, even earlier, the SMR had succeeded in producing domestic passenger and freight cars, for which the manufacturing technology was not complicated. On the other hand, Takahashi noted that light rail was transferred to heavy rail with the improvement of the track section, which duplicated the increase in transport volume. However, "the increase of operation frequency and axle load" caused frequent rail damage. On the other hand, vehicle production and repair in the railway workshops were not fully analyzed in this study. The transfer of railway technology should be related to the operation method implemented by the SMR and to the track improvement they managed to achieve.

Additionally, as Takahashi interestingly noted, the scientific management method was introduced in the company, especially in railway workshops, which improved its efficiency in the 1920s. However, the effect of the scientific management method in railway workshops has not been addressed quantitatively. Bugil Bae (2002) showed that the scientific management of Frederick Winslow Taylor was introduced in the Shahekou railway workshops to improve "efficiency promotion" through the establishment of an Efficiency Office (in 1924) in the SMR. Later, in 1928, a Manchurian Efficiency Research Society, supported by the SMR, was established, leading to the creation of the Manchurian Efficiency Association (1939). This study considers efficiency promotion from the viewpoint of management history, even though it does not verify it for a particular field, such as railway workshops.

Per Minoru Sawai (1998), the technology of railway vehicles was developed inside the SMR. New vehicle production for the SMR was implemented under its strict supervision within the Machinery Section of the SMR, in collaboration with the Shahekou Workshop and The Railway Technology Institute. The Machinery Section provided the design for outsourcing parts to each factory, where controllers of the SMR supervised the production process and product inspection. Yoshihiko Sato (2002) regarded the "Asia Express" train service as a process innovation and noted that the Railway Technology Institute experienced a learning process before the implementation of the plan, which greatly contributed to its realization, enhancing the level of Japanese domestic technology. There was sufficient accumulation of knowledge to devise "the bullet train plan." Nonetheless, like other researchers, Sato did not discuss the characteristics of the SMR in terms of railway technology because his emphasis was on the operation of a specific train, the Asia Express. He did not address how the research results of the Railway Technology Institute were implemented in railway management. In addition, he explained the gap between the year 1930, when the number of reports reached its peak, and the year 1934, when the Asia Express began operating as a phenomenon that could also be seen in the Tokaido Shinkansen. He argued that "the accumulation of knowledge and the progress of tests" were similar between the two railways, but this evaluation ignores the historical fact that research dissemination was limited during the war and the Railway Technology Institute had no choice but to cooperate with both, inside and outside the SMR.[1] Additionally, the researcher who is known for having analyzed the post-war technology transfer of the SMR is Nagami Toshiaki (2003). He noted that, because railway technology was not immediately inherited by the Chinese after the end of WWI, and due to the use of Japanese engineers by the Chinese Communist Party, the employment of Kuomintang engineers, as well as Soviet engineers, was necessary to fill the technical vacuum. How can the technical asymmetry between ethnic groups, which translated into the asymmetry between the high railway technology of the SMR and the technical stagnation of Chinese employees, be explained from a historical point of view?

This study focuses on railway construction and vehicle provision as an indicator of railway technology and considers the overall evolution of the railway industry in Northeast China. The remainder of this paper is organized as follows. In Sect. 4.1, we consider railway construction and improvements, we describe the features of the Korean and Manchurian railway, and we compare them with the JNR. We discuss the production and repair of vehicles in Sect. 4.2. Section 4.3 examines the significance of the development of the railway industry based on the world's technical levels and considers the postwar period from the point of view of Chinese employees. Section 4.4 describes the wartime type of technology, and the last section provides our concluding remarks.

4.2 Establishment of the SMR and Introduction of New Technology

After the Russo-Japanese War, the SMR was established on the Supervision Office of Field Railway, and it started its railway business in April 1907 (Minami Manshū Tetsudō 1919, 148–375). The SMR accounted for a total of 1,150.8 km: The Southern branch lines accounted for 695.2 km between Dalian and Mengjiatun, on the Chinese Eastern Railway, and the Anfeng light railway accounted for 296.4 km between Andong and Fengtian (i.e., Mukden). The SMR not only managed a coal mine, water transportation, electricity, a warehouse, civil engineering, as well as the railway business, but also controlled the administration of the whole area, which became the core of Japanese influence on mainland China. The SMR was referred to as the Japanese version of the "East India Company." In particular, even though the SMR managed the construction of Manchukuo as a comprehensive developer after the Manchurian Incident, later its business scope was reduced to the

[1]The Technology Institute, after the Manchurian Incident, "made a great stride to enlarge research facilities and doubled staff members. Although the number of research articles was not high, this did not mean the stagnation of the research activities. However, the studies of the institute were related to wartime characteristics, which were not made public, and the institute had to cooperate with the Central Institute and railway work-site operations and affiliated companies in terms of design, construction, and implementation" (Mantetsu Tetsukenkai 1990, 31).

administration of the railway business and the reorganization of coal mines. The railway business, though, was still the key driver of the whole business.

However, when the railway business began running, the narrow gauge of the Anfeng Light Railway was 762 mm and the gauge of the main line of the SMR was 1,067 mm, which was much narrower than the 1,435 mm of the international standard gauge. To reinforce the weak transportation capacity and take advantage of the network with other railways, the gauge-widening project of the Dalian-Changchun main line was initiated in May 1907 and completed in May 1908. Later, the gauge-widening project of the Anfeng line was also carried out, from August 1909 to November 1911 (Minami Manshū Tetsudō 1937, 70–75). A double-track construction of the main line was also implemented; a double-track section was open to traffic from Dalian to Sujiatun in October 1909, and it was extended to Mukden in November 1918, to Tieyan in November 1921, and to Changchun in September 1934 (Minami Manshū Tetsudō 1938, 179–192). In addition, the route to the Fushun coal mine was transformed into a double track in 1922, and an electric train began operation in the Fushun coal mine in 1927. However, after the railway capital stock of the SMR increased rapidly from its establishment to the early 1910s, it stagnated (Lim 2013).

The number of railway vehicles managed by the Supervision Office of Field Railway was 220 locomotives, 157 passenger cars, and 3,813 freight cars suitable for the narrow-gauge cars of the main line, as well as 82 light locomotives and 651 freight cars for the Anfeng line. The narrow gauge and the light railway vehicles were sent back to Japan following the standard-gauged reconstruction, and newly manufactured standard-gauged vehicles were introduced. However, 205 locomotives, 95 passenger cars, and 2,090 freight cars were imported from the United States and Japan. Standard-gauged vehicles were received in August 1907 and were assembled to be operated for the first time on the Lushun line in December of the same year. However, railway vehicles were increasingly deployed (Fig. 4.1).

The steam locomotives purchased to replace narrow-gauged vehicles were Amei type passenger locomotives and Davui type freight locomotives manufactured in the US. Although these locomotives were of a "great appearance" at that time, they were soon reassigned from the main line to the branch line of Lushun and Yingkou, due to enlargement of the locomotive. The SMR imported various types of locomotives, including the Paci type of passenger locomotive, the Soli type of freight locomotive, and the Dabu type of shunting locomotive, following track improvements. In addition to American locomotives, the SMR imported the British Sorisa type for the first time in 1912, when "the company was faced with the important choice between the American type and British type for the railway vehicle of Manchuria and had to decide its policy" (Yoshino 1939). The evaluation of the SMR was that "although the British type was somewhat better in engine efficiency, it was less efficient in terms of the structure of the engine locomotive and maintenance than the American type. Because there were advantages and disadvantages respectively, it was difficult to discriminate between merits and demerits." The policy of the SMR was to return to the adoption of the American type of

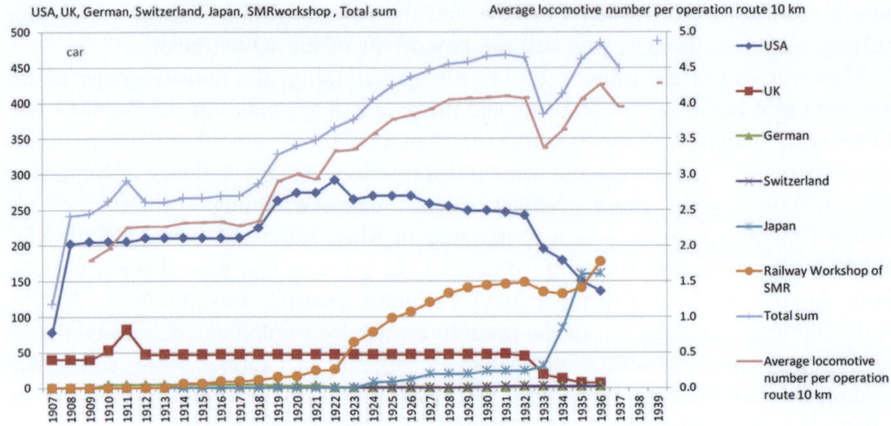

Fig. 4.1 Locomotives of the company-owned lines, SMR by production site. *Source* Minami Manshū Tetsudō (each year). *Note* Due to limitations of the historical data, railway workshop includes only the locomotives assembled in a workshop

locomotive, and the SMR authorities refrained from additional purchases of British-made locomotives, importing many locomotives from the United States.

In the case of passenger cars and freight cars, which do not require high technical standards compared with locomotives and rail motorcars, local manufacturing and self-production were soon promoted. First, even though only 95 passenger cars of six types, such as Ine1, C 3, Rone 2, Ronete 2, Ha 9, Deu 1, and Deu 3 were imported from the US when the SMR was established, the supply of passenger cars became autonomous through the production of the SMR's railway workshops and vehicles manufactured in Japan. As Fig. 4.2 shows, the introduction of new passenger and freight cars continued, following the increase in traffic volume. Regarding the supply of freight cars, 860 boxcars, 1,030 open wagons, and 200 flat-wagons, for a total of 2,090 cars, were supplied by American Car and Foundry Co., Kisha Sezo Co., Amano factory Co., and Nippon Sharyo Co. when the gauge-widening project was implemented. Railway bogies were all American-style. Afterward, freight cars were supplied by the Pullman Company, Ralston Steel Car Co., Ltd. in the US, and the Metropolitan Corporation, Birmingham Rail & Locomotive in the UK, but local manufacturing was rapidly promoted.

The organization responsible for the repair and assembly of these imported vehicles was a railway workshop. The SMR took over the entire site and factory equipment from the Supervision Office of Field Railway and installed railway workshops located in Dalian, Liaoyang, Gongzhouling, and Andong, in April 1907 (Mantetsu Shainkai 1929). At that time, the Dalian workshop was installed in a small shack within the premises of the station, where no machinery and equipment was normally allowed. For this reason, a new workshop was necessary, and a construction plan was established. The SMR began the Shahekou workshop construction in July 1908 and gradually moved the factory equipment, with the

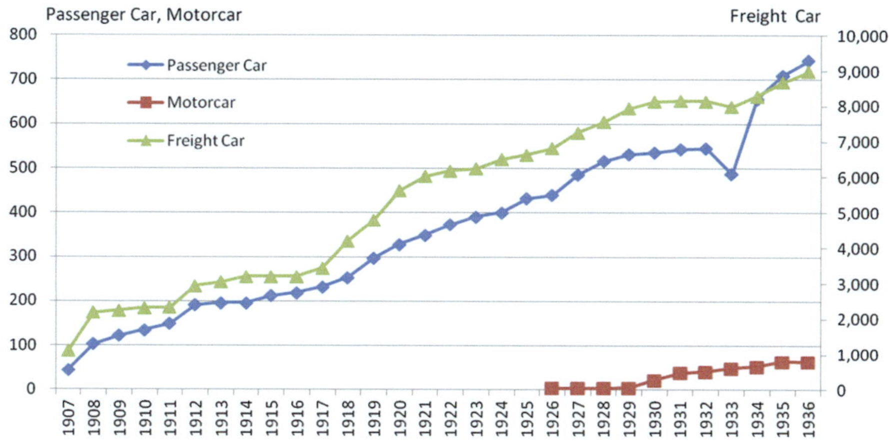

Fig. 4.2 Passenger car, Motorcar, and Freight car of the company-owned lines, SMR (unit: car). *Source* Minami Manshū Tetsudō (each year)

progress of the workshop construction, to complete the whole operation in August 1911. While the repair workshop for locomotives was located in the center of all routes, to reduce the transport cost of returning from an engine depot to a repair workshop, the repair workshops for passenger and freight cars were located near the terminus where passengers and freights got on and off a train. Because they were many workshops, two workshops for passenger and freight cars, Gongzhouling and Andong, were closed and only two workshops, Shahekou and Liaoyang, remained.

The SMR not only carried out an extension of the workplace and the improvement of various machinery and equipment as described above, but also opened a Railway Apprentice Training Center (Mechanic Training Center) in 1909 to train mid-level engineers and enhance repair skills. When the workshop apprentice regulation was enacted in 1929, the SMR started to educate the Chinese separately, and the Railway Apprentice Training Center trained only the Japanese. In addition, an incentive system to encourage employees' work efficiency was devised. Although a daily wage was paid based on working hours in railway workshops, a Rowan premium bonus work system was introduced in September 1911, and a supplement was paid for premium work hours. The Rowan premium work hour was calculated by B × (A-B) ÷ A, in which A and B implied a decided work hour for completion and a real work hour for completion, respectively.

Unlike the Japanese National Railways, the SMR's workshops were not limited to the repair of railway vehicles and the production of railway facilities but also engaged in the production of various types of machinery and non-railway equipment for business diversification, as well as in the production of vehicles. The workload of railway workshops increased following the outbreak of WWI. The Shahekou workshop oversaw the production of various machines and materials for the construction of the Anshan steel plant in 1916, and, thus, the workload of railway workshops increased dramatically. Since it was difficult to rely on the scheduled completion period with the workshop equipment of those days, the workshop expanded its electrical equipment, and extended the workplace for casting, rivet joint, and finishing, in 1917. However, when WWI ended, the price of iron and steel slumped, and the Anshan steel plant was forced to stop its expansion project. A policy of business reduction was inevitable for the Shahekou workshop, too, which engaged only in the repair and manufacture of railway cars.

As described above, even though the SMR began railway operation in mainland China, the production capacity of locomotives was not yet developed on the Japanese side, and technology improvements continued with the American locomotive as a baseline model. Eventually, the SMR achieved domestic production and self-development, as described in the next section.

4.3 Technology Accumulation and Self-development

Even if transport shortages became significant during WWI, the supply of vehicles was negatively affected by the price of capital goods to be imported from abroad, in particular from the US. An active railway investment was made after the end of WWI. The increase of rail weight began in 1919, and existing rails of 60 lb, 64 lb, and 80 lb were replaced with new 100 lb rails. The heavy rail section, where the increase of rail weight advanced, passed 1,110 km in length in September 1926 and reached 1,783 km in March 1936 (Minami Manshū Tetsudō 1928, 185–227). At the same time, the number of sleepers per rail increased from 14–15 to 18–21 with the strengthening of the track bed. The reduction of the slope and curve radius, straightening of the line, and reinforcement of bridges were also carried out (Fig. 4.3).

These measures became necessary following the weight increase and the high speed of trains, and the SMR installed additional blockers, especially composite blockers, as well as various railroad signals and switch indicators to increase the frequency of train operation. Automatic blockers were also installed in the main line (Minami Manshū Tetsudō 1938, 212–224). The SMR then added stations and platforms to improve railroad transport in wide areas along the railway. As the increase of frequency in train operations naturally required accurate and frequent contacts between the locomotive engineer and the station and railway office, the SMR installed additional telephones and telephone lines (Minami Manshū Tetsudō 1928, 194–201, 1937, 90–99).

In addition, the SMR introduced large engine locomotives and increased the number of connected cars per train. For example, in the case of a freight train, the number of connected freight cars increased from 22 cars in 1912 to about 40 cars in 1936. At the same time, the SMR increased engine depots in charge of locomotive operations and introduced various equipment, such as water taps, ash pits, and turntables, to enhance locomotive operation capacity. As for railroad cars, the SMR imported many of the Mikai type, the Mikani type, and the Dekai type of locomotives for freight trains, especially in 1920 and 1924, and "possessed presentable

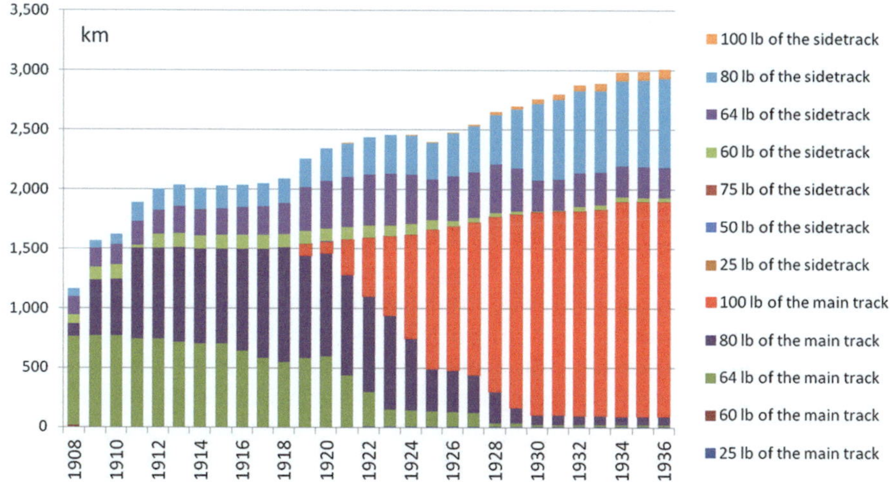

Fig. 4.3 Rail weight of the company-owned lines, SMR. *Source* Minami Manshū Tetsudō (each year)

railroad cars in comparison with first-class JNR and other railway companies" (Yoshino 1939). As shown in Fig. 4.1, the American type of engine locomotives became the mainstream, except for a small portion of Swiss and German-made locomotives, with a strong American influence on the background of the SMR's railway technology.

Even though the initial technical level of the SMR was nothing but the assembly of purchased vehicles, the SMR devised a new production project in 1912, remodeling the design of American-made locomotives to manufacture a Sorisi type of freight locomotive in 1914. Self-production progressed with the enhancement of a domestic production policy for engine locomotives, the SMR began producing a Pacini type of passenger locomotive in 1916, and the SMR began producing an electric locomotive for the Fushun coal mine, which was the first attempt for the company, in 1917. Railway workshops not only produced various types of locomotives but also began producing for outside companies. However, they only produced strong engine locomotives (i.e., the Deka type of freight locomotive and the Pacisi type of passenger locomotive) after 1921; a newly manufactured Mikai type of freight locomotive with three-cylinders, unusual for the railway vehicle industry in Japan, was produced from 1926 to 1931 and acquired a favorable reputation (Minami Manshū Tetsudō, 1938). Given the high speed of the steam locomotive in the rest of the world, the Paciko type of locomotive was manufactured for the express train "Dove," connecting Dalian and Changchun in 1927. It was "the first locomotive manufactured with the SMR's own design," and it showed various improvements, such as the installation of a combustion chamber in the engine to enhance the efficiency of the coal burn; it had a good reputation thanks to the operation efficiency of the "Dove Express." The Paciko type of locomotive contributed greatly to the long-distance service between Dalian and Changchun and held "the premier position of passenger locomotives until the advent of Pacina type locomotives" (Yoshino 1939). In 1931, a railway workshop accepted an order from Qingdao–Jinan Railway, and manufactured four Mikado type locomotives, exploiting the technology accumulated by the SMR, which received high praise for "its swiftness and the efficiency of the coal combustion is a wonder in this industry."

On the other hand, an important step in the improvement of passenger cars was the material of the car body. Even though it was initially wooden, with time, a higher solidity of the car body was requested, following train extension and an increase in speed. Passenger cars began being produced in steel frame wood in 1917, and an all-steel car body was adopted in 1926, earlier than in other railways. At the same time, the roof of the car changed from a high-window type to a Monitor type, and round-roofed car were introduced to reduce train running resistance and manufacturing cost as much as possible. However, as the introduction of the all-steel car body brought about a relevant weight increase, the SMR devised new measures such as a welding design and the reduction of the cross section and metal fittings.

For the railway bogie, an important component of passenger cars, Pullman, American Car, and Foundry types of cars were imported. These were also American types of balance beam bogies. The SMR adopted the C-axis 36-wheel type, based on the Pullman type bogie, which consisted of two 4-wheel and 6-wheel types. In contrast, because American Car and the Foundry type, which had the B-axis 33-wheel and seven-foot wheelbase, were unsuitable for high speed and had difficulties in replenishing wheels, as wheels were chill hardening, these types were not adopted, except for special cars. As the Pullman type bogie needed strengthening of the braking force due to the high speed of the train service, the frame timber type became difficult to procure, and the steel twin braking type was introduced in 1929. With this improvement, for the first time a roller bearing was chosen, and this experience was used for manufacturing high-speed deluxe cars for the "Asia" train. As a coupling device, although the three-lever type was adopted to cope with the increase in the volume of passenger transport, it was soon replaced by the Shibata type of coupler, in 1928, and by a wheel ring spring type in 1930. As a braking device, the American Westinghouse M type was adopted.

Regarding freight cars, the body of boxcars was wooden at first and was changed to steel in 1924 to cope with the increase in speed and the extension of train units. A 60-ton steel boxcar was manufactured in 1927 and a Yasa type of boxcar, equipped with a lifting platform and a toilet, was produced in 1928; it was used for coolie transport. An open car, steel table frame, and pillar were introduced in 1921, and partial improvement was added to it. The Muni type bottom dump open car was manufactured for ballast spraying in 1931, and an improvement was added to the bottom in 1936. In addition, various types of wagons, such as hopper cars, coal cars, refrigerator cars, and flat cars were manufactured in accordance with the specific needs of the cargo, and many improvements, including a steel car body, were added to improve strength, reduce weight, and simplify production.

How was this original design of the SMR adopted? Testing and research of the Railway Technology Institute, the design technology of the Machinery Section, and the manufacturing technology of the Railway Workshop contributed to achieving a vehicle production featuring the SMR's own design (Ichihara et al. 1971, 263–271). The Technology Institute, which implemented the applied testing of equipment and materials for the company and led to the overall improvement of technical facilities, was installed in the Shahekou workshop in March 1922, and its facilities were gradually expanded, especially after the construction of a new main building in 1928 (Mantetsu Tetsukenkai 1990). Research activities focusing on machinery, vehicles, rail, civil engineering, and electricity continued, and Thursday meetings were established in August 1925, to perform research presentations and literature discussions, sharing and spreading knowledge. The research results provided useful insights for the maintenance of railway facilities and for the increase of railway vehicles. This institute changed to the Physical Testing Laboratory in June 1930, and in December 1931, the Shahekou Institute of Central Laboratory Institute was separated from the Laboratory, to be integrated into the Railway Institute as a Dalian Branch Office, in March 1937, and then expanded to the Railway Technology Institute in April 1939. The Railway Technology Institute made every effort to test, improve quality, and develop new technologies related to a wide range of railway facilities and vehicles, to support the SMR technologically. Furthermore, while the design of new cars had so far been developed by the Machinery Section, engineers graduated from engineering colleges, and universities gradually took control of the design area. In the case of the SMR, there were many graduates from South Manchuria Technical College and Lushun Institute of Technology (Amano, 2012, 34-48). These technicians were not restricted to a particular department, and they were relocated to institutes and workshops, through which the cooperation among the three agencies continued and the commercialization of technology constantly improved.

This study examines the railway workshop in charge of the production of vehicles. First, the accumulation of technology did not start with production, but it became possible through the assembly of new vehicles and routine repair works. In this respect, it is worth noticing that a routing system was introduced into the repair work. Due to the lack of historical data, there is uncertainty regarding the routing system in the Shahekou workshop. However, in the Liaoyang workshop, the routing system was introduced as a new repairing method in April 1921 (Chūgai Sangyō Chōsakai 1923, 37–40). This system was introduced following American railway companies by the JNR in the first half of the 1910s and became the key repair technology for the rationalization of locomotive repair work (Lim 2010, 271–298). Once accomplished in Japan, this repair technology was also disseminated to the SMR, one of the colonial railways.

The new technology operated as follows: (i) Workshops investigated the state of the scheduled car for entrance, as preparatory work before the entrance of the locomotive, and prepared material for replacement and processing for the day of entrance. (ii) Each workplace prepared status quo tables for each locomotive and promoted a specific preparatory work. (iii) Workshops asked the engine depot to submit a written report for the scheduled locomotives for the next month's entrance to investigate the repair parts required, compare the report with status quo tables, and send a technician if necessary. (iv) Workshops wrote status quo tables based on the entrance-exit schedule table of the regulation of transportation office and submitted a table to each worker involved. (v) Workshops determined the average days required for canning and assembly and compiled a master card (or master schedule).

In the post-entrance phase: (i) After a workshop dismantled a locomotive and sent a boiler to the canning workplace, it wrote an inspection report, assessed the degree of the repair, decided the scheduled completion date according to each repair grade by the master card, and notified the employees. (ii) All employees involved, including the locomotive workplace chief and foremen, conducted a meeting about the many dismantled mechanical parts and filled repair points on the inspection sheet, to submit to the team leader and promote the repair work. (iii) Without special repair work, a workshop filled the scheduled date sheets of each work completion by the master card of each workplace, based on the scheduled date of boiler repair completion, and submitted it to each workplace chief. (iv) Each workplace tried to finish the workload within the time limit. Thus, it was possible to reduce the work time by more than forty percent from April to September 1921. This routing system became the basis for vehicle repair, but it should be noted that spot repairs remained for some work (Minami Manshū Tetsudō 1928, 386–417; Asō 1933, 5).

The efficiency enhancement of the SMR also led to the full-scale introduction of the Taylor system, which did itself proud in the world. Scientific management methods were introduced in Manchuria by Shoji Ishihara of the Lushun Engineering School, renamed later as the Lushun Institute of Technology (Kaise 1931; Ōnaka 1940). Under his influence, Tsugio Ouchi entered the SMR at the end of 1921 and wrote a report titled "The wage system of the Shahekou workshop." The technical committee asked him to assess the efficiency of workshops in 1924, and he reported this private opinion about their efficiency. In a second meeting held in November 1924, Housaku Akita, chief of Dalian Engine Depot, proposed "the establishment of an efficiency institute within the company," and Motoki Yamazaki, who attended the meeting as secretary of the correspondence department, supported it. Later, Kyōmeikai, who organized mid-level employees to study the internal and external affairs in that period, also took up the efficiency institute case in the meeting of January 22, 1925, and issued a positive evaluation. On the next day, Motoki Yamazaki drafted a "case related to the installation of efficiency department," which received a settlement as a "case about the Installation of an efficiency subsection" on April 1, 1925. An efficiency subsection was installed in the president's office as a result. The efficiency promotion policy of the SMR led to the establishment of the Manchuria Management Association, in 1939.

The team leader system was replaced by the functional foreman system of Taylor in railway workshops from 1925 to 1926. The team leader system was the workplace regime in which team leaders or foremen oversaw tens of workers under each workplace chief and were engaged in repair work (Mantetsu Shainkai 1929; Takahashi 1931). Even though these team leaders and foremen generalized every type of work, such as work planning, preparation, supervision, inspection, and work distribution, their burden was inevitably so large that they were confronted with tremendous difficulties and defects (Takahashi 1940, 43). For example, a person with a talent to supervise subordinates was not good in the control of the repair plan and works. To avoid such inefficiencies, the scientific management method of Taylor was introduced, and the work organization was restructured. The functional system was carried out; each subsection, such as a plan subsection (repair work method decision, work distribution, and deadline determination), a preparation subsection (preparation and distribution of material tools), a guidance promotion subsection (guidance for workers and progress of work), and a maintenance subsection (maintenance of machinery, tools, and buildings) became responsible for professional work. Indirectly engaged employees increased as compared to the conventional team leader system, but the efficiency improvement realized by the new institution was carried out without increasing production cost. The introduction of the new system was made possible by many excellent foremen who supported the new Taylor system with their skills in each site.

It is worth noticing that the development of the workshop system reflected the ethnic hierarchy. In the Shahekou workshop, the working range of the Chinese-Japanese ethnic groups was different; while difficult or high-quality work was almost exclusively assigned to Japanese workers, only simple or low-quality work was reserved for Chinese workers. Both ethnic groups engaged in middle-level work. For example, regarding the assembly workplace, some tasks were seen as impossibly hard for Chinese workers to carry out, and even the most outstanding Chinese workers, who oversaw small-parts production, were never assigned: (i) work requiring the examination and comprehension of drawings; (ii) work requiring intellectual consideration; (iii) work requiring high precision; and (iv) work requiring supervision to generalize various aspects of the process. This was a general evaluation applied in most workplaces, and such skill differences were also reflected in the position gap. In the lathe workplace, even the kind of machinery to be used was different[2] across ethnic groups. In terms of skills and abilities, Japanese workers were about 20% superior to Chinese workers. Moreover, workshop authorities categorized Chinese employees as follows: (i) the Chinese did not feel a strong sense of responsibility; (ii) they did not show endurance in performing to the end; and (iii) they did not understand the language well. It was also noted that "if the scientific management method, that is, the system currently implemented in the Shahekou workshop, was fully embraced and all work standardized, and if craftsmen worked under the guidance of planning subsections, there would be no difference between the abilities of workers belonging to different ethnic groups" (Rinji Keizai Chōsa Iin-kai 1929).

[2]While the machine tools used only by the Japanese included the milling machine, big lathe, planer, bevel gear generator, gear bobber, and universal grinder, the machine tools used only by the Chinese were the drilling machine, grinder, wheel lathe, and screwing lathe. The machine tools used by both Japanese and Chinese workers were the small lathe, small planer, crank planer, and turret lathe (Rinji Keizai Chōsa Iin-kai 1929).

In the Shahekou workshop, which aimed at the localization of the scientific management method, "the diligent organized week" movement was carried out and "duties encouragement, efficiency enhancement, workplace cleaning, material sorting, [and] equipment repair" were emphasized (Mantetsu Shainkai 1929; Takahashi 1931). Various committees, such as the committees responsible for Monday meetings, Wednesday meetings, and Saturday meetings; the standard practice committee; the office work investigation committee; and the committee responsible for disaster prevention meetings were introduced to discuss savings in fuel, materials, and consumables, and to help achieve the improvement of the working method, accident prevention, and efficiency enhancement. Furthermore, the "meeting" between the workforce and the management, with 57 members elected from the workers together with executives, was held in spring and fall, twice a year. In the meeting, the improvement of factory work and workers' expectations were discussed. Workers of all ethnic groups experienced difficulties under the influence of Yuaikai; even the Communist Party experienced difficulties, with at least one strike every year after the outbreak of WWI (Minami Manshū Tetsudō 1929). The stabilization of the labor-management relationship through the "meeting," which functioned as a kind of "factory committee," helped improve work[3] in Japan.

When the name of the Shahekou workshop was changed to Dalian workshop, in 1928, the SMR abolished some departments, such as general affairs, supervision, ironworkers, and woodworking, and placed a deputy manager and a chief engineer under a workshop manager to support the manager. They put the workplaces under the direct control of a manager to strengthen the work system of railway workshops (Minami Manshū Tetsudō 1938). In addition, even though "the outcome of workshops" was manufactured products respecting basic cost principles, in 1928 workshop authorities emphasized efficiency enhancement as part of the scientific management method and changed the actual cost principles to the contract system, creating "tension" among workers (Minami Manshū Tetsudō 1938). In the workshop for recycled products, the Halsey Premium work system (Halsey premium days = (A-B)÷2) was introduced only for a part of the work, in June 1933. While half of the premium days were paid in addition to the basic salary of individual workers, the other half were paid based on "the average salary of all workers, by Chinese-Japanese ethnic groups, respectively." As the company turned to budget austerity during the Showa Depression, the Liaoyang workshop was closed and merged with the Dalian workshop in January 1930. A sharp decrease in workload resulted in excess work capacity, and a temporary shutdown and cut-back in operations was implemented as a countermeasure.

[3]"When it comes to workers' thought, because the thought of both Japanese and Chinese is now moderate, and communication and mutual understanding with the workshop executives were good enough, both ethnic groups remain very calm. However, our workshop, in which many workers participate, should always pay careful attention to workers and continue to consider the employees' issue of welfare, while making an effort to not neglect it." (Mantetsu Shainkai 1929, 10–12).

Progress in work organization shortened the completion time of all work, as shown in Fig. 4.4. Railway workshops carried out not just repair work, but also the production of vehicles and railway facilities. Figure 4.5 shows how the work of railway workshops can be transformed into a real sum of money using a deflator and dividing it by the number of the railway workshops' employees, to estimate the trend of labor productivity. After the beginning of WWI, the output of workshops (i.e., the amount of work employed) increased rapidly, and the number of employees increased to cope with it. However, after the end of the war, as output did not return to its pre-war levels and, in fact, was actually reduced, workshop authorities carried out several layoffs and aimed at streamlining. Because workshop authorities introduced a casual worker system in 1924 and tried to avoid an increase in personnel expenses, many casual workers were hired, largely as a part of management stabilization after the Showa Depression, through the 1930s. Meanwhile, labor productivity greatly improved and continued to increase in the 1920s. Therefore, it seems that the effect of the scientific management of Taylor was large.

Even though both English and American technologies were introduced, the SMR adapted American railway technology to the continent called Manchuria, and achieved its own originality, realizing a high level of productivity.

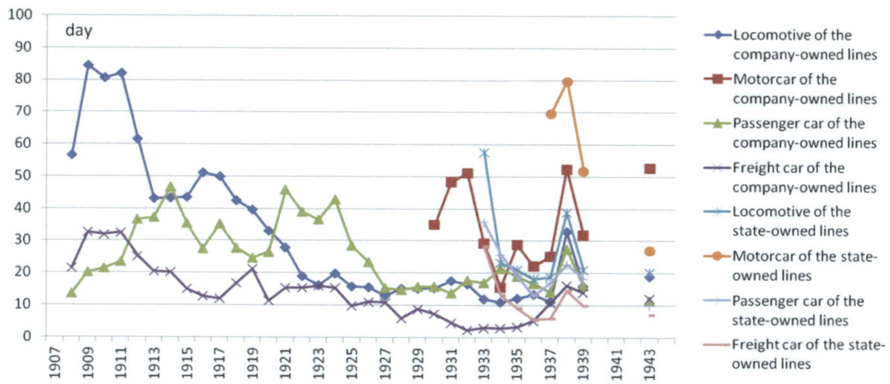

Fig. 4.4 Average number of days needed for repair in railway workshops days. *Source* Minami Manshū Tetsudō (each year) and Minami Manshū Tetsudō (1943). *Note* Displays the average number of general repair and local repair

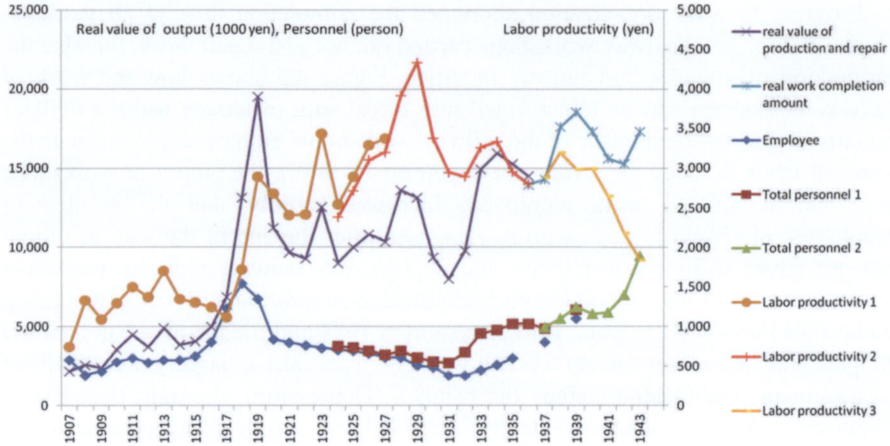

Fig. 4.5 Workloads and labor productivity in the workshops of the company-owned lines, SMR. *Source* Minami Manshū Tetsudō (each year), Minami Manshū Tetsudō (1938, 377), Shomuka (1930), Shiryōka (1942), Minami Manshū Tetsudō (1943), and Ōkawa et al. (1966). *Note* 1. In terms of production value in the workshops of the company-owned lines, because the aggregation method changed in 1936–1937, the production value was calculated as the amount of production and repair, from 1907 to 1936, and as the work completion amount after 1936. 2. Total personnel 1 = employees + temporary workers. Total personnel 2 = employees + non-regular employees. Although a temporary worker system began in 1924, because the number of workers in 1925 was not known, we estimate it in this study by linear interpolation with the same number in 1924 and 1926. In addition, the employee number in 1936, which was also not known, was estimated as the ratio of employees and workers in two following years, 1935 and 1936. 3. Labor productivity 1 = real amount of production and repair ÷ employees. Labor productivity 2 = real amount of production and repair ÷ total personnel 1. Labor productivity 3 = real work completion amount ÷ total personnel 2. 4. The prices index of railway vehicles and related equipment (Ōkawa et al. 1966) was used as a deflator for the railway workshop output

4.4 Management of All Manchurian Railways and Technology Diffusion

After the Manchurian Incident, the SMR took control of the operation and restoration of occupied railways, following the directions of the commander of the Kwantung Army. Thus, its business scope was not limited to existing company-owned lines but extended to all railway routes in Manchuria. In other words, the railways of the Chinese became part of the Manchukuo National Railways (i.e., state-owned lines) and fell under the control of the SMR in February 1933. The Northeast Korea lines and the Korean National Railways of the Hamgyong, the North Province of Colonial Korea, were consigned to the SMR at that same time. The Railway General Office and the North East Korea Railway Management Office were established to operate these railways as external bureaus of the SMR.[4] When the North Manchuria Railway (i.e., the Chinese Eastern Railway) became part of Manchukuo in 1935, this railway also became part of the SMR, constituting state-owned lines.

In part of the State Line, the Railway General Office established an annual plan to maintain and improve the facilities of the State Line, with civil engineering as a top priority project. At first, most projects showed considerable progress; however, several issues, such as the shortage of cement and a delay in the supply of sleepers, occurred. In particular, because facility standards were very different across lines and the degree of deterioration of facilities was high, an Equipment Commission was established in the General Office to judge the situation, considering the specificity and circumstances of the State Line, and to draft an intervention plan after investigating and researching all relevant issues. In other words, the Equipment Committee oversaw the investigation of important issues, such as basic equipment planning, technical planning of new facilities, equipment control, operational planning, equipment standard setup, and the relative implementation plans. In addition, as new routes built by the Railway Construction Bureau, including the Jingtu railway, were placed under the administration of the SMR and simultaneously taken over by the Railway General Office, the state-owned lines continued to expand rapidly.

The SMR unified various standards and institutions of the state-owned lines, with the standards of the company-owned lines as a model, and tried to finally integrate the state-owned lines with the company-owned lines. Therefore, the SMR merged nine railway bureaus (Fengshan, Shenhai, Jihai, Jichou, Jidun, Sitao, Taosuo, Qike, and Huhai) with the four bureaus of Mukden, Changchun, Harbin, and Taonan, based on the fundamental management outline released in April 1934, unified transportation regulations, and made employee treatment across railways uniform. The Transportation Department was installed in the Engine Depot of the Railway General Office, to administer affairs related to special transportation, beginning in November 1934. The Railway General Office decided on a "Tentative Improvement Plan for the Future" in October 1934, and a "State-owned Lines Improvement 10-year Plan," in December 1934, to seek the unification of different facility standards by route while strengthening the existing facilities (Mantetsukai 1986, 236–265). Meanwhile, the expansion of the signal field; the renovation of the garage, railway turntable, passenger platform, and cargo unloading facilities; and the expansion of an open storage area and a sidetrack were achieved. In terms of communications, the Railway General Office installed the Communication Office within a unique organization, responsible for the installation of the garrison telephone line, car dispatch lines, telegraph lines, and telephone relay line for the integrated communication network to cover each route.

[4]Among Northeast Korea Lines, the Western line that was the route between Sansambong and Chongjin, the route between Hoeryong and Sinkaerim, and Chongjin port were returned to the Governor-General of Korea, and the remaining route of 183.3 km between Sangambong and Eungki was lent to the SMR (Senkōkai 1986, 83–84, 87).

The SMR earned the consent of the Kwantung Army and the Government of Japan, and unified all of Manchurian railway management under the office reorganization of October 1936, to establish a Railway Secretariat General in Mukden as a general management agency of the 1,129.1 km of company-owned lines, 7,421.5 km of state-owned lines, and 344.4 km of Northeast Korea Lines, for a total of 8,895 km (Mantetsu Shainkai 1936a, 24; Mantetsu Shainkai 1936b, 18–19; Shigeru 1936, 20–21, 48; Mantetsu Shainkai 1937, 26). Thus, for the unification of railway management of the company-owned lines, state-owned lines, and Northeast Korea lines to become possible, the burning issue was how to introduce unified provisions, such as fares, operations, and construction. Although the state-owned lines and Northeast Korea Lines were integrated as one railway system through the management of the SMR, since they were operated by different standards and management systems, the SMR had to integrate these railway facilities and institutions. Because the integration of the communication systems was urgently needed to supply a unified train service, the Railway Secretariat General unified the types and treatments of telegraph, telephone, and wireless communication to achieve comprehensive development (Mantetsukai 1986, 265).

As the business scope of the SMR was rapidly expanding, the vehicle sector also made great progress. As part of the state-owned lines, a rationalization of railway workshops was also promoted. In addition to taking over the railway workshops of Huanggutun, Changchun, and Shongpu to concentrate the workshop ability on repair work (Tetsudōbu 1932), the SMR devised a plan to supplement the labor force of each workshop and assigned several Japanese instructors from the Dalian railway workshop, the JNR, and others to workshops of state-owned lines; the SMR tried to increase the Chinese workforce in each workshop in the period between 1933 and 1934 (Minami Manshū Tetsudō 1938, 1113). Even though each workshop had a distinct organization, the Railway General Office carried out an organizational change in August 1934, pushing for the unification of all workshops under the office reorganization amendment, in April of the same year. The Railway General Office also extended its account and salary regulation to all workshops in April 1934, established the working hour system for each workshop, and carried out the unification of accounting principles. In April 1935, the SMR amended the wage system of direct workers and adopted the Rowan premium wage plan completely as a reasonable wage system to improve work efficiency and secure the accuracy of production costs.

Existing workshops were reorganized, and the SMR started to enlarge workplaces involved in freight car repair in the Huanggutun Workshop beginning in 1934 and completed the establishment of the wagon lumber and bogie workplace. The SMR then embarked on the construction of the Sankeshu railway workshop in 1935, and since the Harbin workshop was established following the requisition of the North Manchurian Railways (i.e., the Chinese Eastern Railway), the Sankeshu workshop became a workshop specializing in locomotives, contrary to the original plan. The SMR installed apprentice training institutions in every railway workshop beginning in September 1936, to alleviate the qualitative dilution of the labor force and begin the self-training of workers, prevent labor turnover, and foster qualitative improvement of skills and performance. Educated trainees and graduates began being hired as workers in railway workshops. Because of the rationalization of repair work and the unification of vehicle standards, the number of days needed for repair in railway workshops was significantly shortened, as shown in Fig. 4.4.

In terms of car repair, all railway workshops of the SMR were subject to management control by the Railway Secretariat General to obtain a railway unification measure, wipe out the different labor controls of workshops between company-owned lines and state-owned lines, and bring efficiency to workload assignments, the arrangement of facility and personnel, and general task performance (Minami Manshū Tetsudō 1938, 1115). Therefore, a regular workshop manager conference, a workshop task study meeting, and a work routine meeting were held to improve the overall renewal of the business. Materials for the workshop were managed as commissioned inventories to improve the convenience of material preparation and management.

After the Manchurian Incident, as traffic volume surged, railway workshops also became "extremely busy" for repair, remodeling, and orders both inside and outside the company, especially orders from the army, such as armored trains and armored automobiles. When the Railway General Office and the Railway Construction Bureau were established in April 1933, workshops were getting flooded with orders from new agencies, as well as the orders described above. In addition, the relocation of many workers to the Railway General Office and other workshops, following the requests of their military relationships, caused a sharp decrease in repair capacity, and workshops of the SMR had no option but to hire workers from the JNR and allow overtime, holiday work, or night work, while expanding equipment to enhance overall productivity.

When Manchukuo was founded, the SMR tried to operate a world-class passenger express for propaganda purposes. Was the Asia Express itself a symbol of Japanese politics, as well as a proof of the technological capacity of the SMR? It matched well with the establishment of the new capital (Shinkyō). The super express manufacture project was proposed in the Railway Section of the SMR in 1932, and the Machinery Department of the Railway Section secured the cooperation of the Railway Institute and carried out the necessary speed tests with the Paciko type of locomotive in July 1933 (Ichihara 2010; Amano 2012). The results led to the approval of the Asia Express project in the executive meeting of August 23, 1933, which established: (i) eight hours and 30 min' operation between Dalian and Changchun; (ii) the installation of air conditioning equipment in all cars; (iii) the beginning of operation at the time of timetable revision in October 1934; and (iv) the train composition of four cars. Later, the product of the Carrier Engineering Corporation was chosen for the air-conditioning system, and the SMR began installation work with the help of employees sent from the manufacturer, to better adapt the basic design of Carrier Engineering Co. to the Manchuria climate. While the passenger car was designed and manufactured by Hiroshi Kojima and others, the design of the Pacina type of traction locomotive was developed by Shintaro Yoshino.

Yoshino's newly designed Paciko Type of locomotive was based on the SMR's original locomotive designed in 1927. Taking advantage of the SMR's experience, he drew up a blueprint for the main part, based on the Paciko from January 1934, and assigned "parallel designs" to 17 draftspersons. Due to the time limits to begin train operation, the Pacific type of locomotive (4-6-2) continued to be adopted, and the new locomotive design was completed in less than four months. Because the wheel diameter was designed to reach 2000 mm to obtain maximum performance, the wide firebox had to secure the high vapor pressure needed for big wheels and an automatic stoker was attached.[5] To cope with the increase in weight due to the adoption of streamlined design, weight saving measures, such as an aluminum plate, aluminum castings, and duralumin were devised to replace steel parts, and welding was used instead of rivet joints. For the first time, a tender was equipped with roller bearings that allowed a no-touch long-distant drive. Pacina type 3 locomotives were manufactured by the Shahekou workshop of the SMR, eight locomotives were manufactured by the Hyogo factory of Kawasaki Sharo Co., and one locomotive was added in 1936.

The passenger cars of the Asia Express became improved with soundproof wheels. Although the SMR adopted a Westinghouse M-type braking device and improved the braking device to LN type after 1932, this improvement was limited to the Asia Express. The coaches of the Asia Express were not only streamlined but also accessorized with many new technologies, such as air conditioning equipment, which was not even out of the testing period in the United States. When the Asia Express passenger train began to operate, special steel was adopted for the main part of the passenger car, and a light alloy was used for the metal fittings. The entrance that used to be in the center was also changed to double doors. A cork powder layer and moisture-proof felt layer were applied for roof and heat insulation, and the entire surface was equipped with a soundproof structure, with flannel.

Thus, the Asia Express could achieve 120 km per hour and, with its sophisticated appearance, it became "a world-class railway" as "a symbol of Manchurian culture" (Yoshino 1939, 34–35; Minami Manshū Tetsudō 1938). The operation of the Asia Express required a change of the cant of the railway track and brought new improvements to the lower part of the line. At the same time, as shown in Table 4.1, the Company Line of the SMR recorded very high productivity compared to other railways. However, this study emphasizes that the operation of the Asia Express was never helpful in terms of profitability for the company management. "Because the net operation cost of the Asia Express between Dalian and Changchun (excluding track maintenance costs and labor costs) needs about two thousand yen for one operation, it was being operated at the expense of the SMR, ignoring profitability" (Ueda 1935, 220–221). This shows that the high profitability of the SMR's business supported the operation of the Asia Express.

[5]Pacina type locomotives had a full-length of 25.7 m, height of 4.8 m, and full width of 3.2 m, with a weight including a tender when filled with coal and water of 203 tons, wheel diameter of 2000 mm, and wheel weight of 24 tons. As a result, the Asia Express train had a record length of 174 m, passenger weight of 340.5 tons, and passenger capacity of 292 persons.

Table 4.1 Operation and labor productivity of foreign railways in the Mid-1930s

	Operation route (km)	The number of employee per operation route 1 km (Person/km)	Revenue per operation route 1 km (10,000 yen/km)	Operating ratio (%)	Labor productivity ([passenger-km + ton-km]/ Person)
Japanese National Railways	16,427	12.8	2.8	60.6	171,456
Company lines, SMR	1,129	27.9	11.9	37.6	251,206
German National Railways	54,375	12.1	8.9	88.1	161,849
Paris-Lyon-Mediterranean Railway	9,894	10.9	5.6	105.2	119,991
French Eastern Railway	5,132	11.0	5.3	108.1	151,172
French National Railways	9,162	8.3	3.5	120.8	97,305
French Alsace-Lorraine Railway	2,300	14.3	5.9	111.1	119,635
Belgian National Railways	4,849	17.1	4.8	101.3	131,961
Dutch National Railways	3,421	9.3	5.5	95.7	
Danish National Railways	2,514	7.8	3.0	100.0	104,721
Swedish National Railways	7,455	3.6	2.0	82.0	160,000
Norway National Railways	3,623	4.1	1.5	102.8	81,933
Spanish Northern Railway	3,803	10.8	3.6	79.0	80,659
Spanish Madrid-Zaragoza Railway	3,670	9.3	3.0	85.6	75,731
Austria Federal Railways	5,801	9.6	4.2	111.0	90,757
Yugoslavia National Railways	9,352	7.5	1.5	104.3	67,358
Polish National Railways	17,895	9.7	2.8	83.9	132,605
Bulgaria National Railways	3,220	5.2	1.2	87.5	69,702
South Africa Federal Railways	21,276	3.4	2.1	65.5	124,608

Source Tetsudō Sōkyoku (1938)

On the other hand, the greatest technical innovation was the introduction of Mikasi type locomotives, evaluated to technically achieve "a remarkable leap," using a high-pressure and high-temperature steam, which was used efficiently, and realizing a significant coal and water consumption reduction (Yoshiada 1939, 34–35). The Mikako type of freight locomotive began being manufactured in 1935 and quickly overcame the disadvantages of the Mikani type, equipped with a three-cylinder type engine and various measures, including an increase in engine capacity for high speed. It became "a great success" based on the SMR's original design, and a leading example of the freight locomotives of the time.

The Dabusa type of small, fast locomotive deserves separate consideration. This locomotive was manufactured by "the best of modern science" based on the experience and knowledge that the SMR obtained in mainland China. The engine of the SMR was exceptionally excellent, far superior to those of conventional locomotives. The SMR manufactured the streamlined Paciha type locomotives in 1936, achieving various improvements, such as the use of roller bearings on all axes, together with the Dabusa type, and the locomotives were run as engines of ordinary express trains in company-owned lines. In the same year, workshops manufactured Deka type locomotives for the Longhai railway, which was designed to be lilting and suitable for mainland China, like Mikado type freight locomotives, which were produced for the Qingdao-Jinan railway in 1931.

The SMR also tried to reduce the operation interval, supporting the development of internal combustion engine vehicles abroad and the production of "innovative" high-speed diesel electric motorcars equipped with 500-horsepower diesel engines, made by Sulzer Ltd. and Niigata Engineering Co., Ltd., and organizing six streamlined trains, with four cars each: one diesel motorcar with a luggage compartment and three trail cars, following the SMR's control method in 1935. Since the urban and intercity traffic volume increased, a duplicate operation of motorcars and a multiple unit control were devised, and the SMR workshops adopted the newest liquid type for power transmission devices and produced Keha 6 and Keha 7, equipped with a combination of fluid coupling and Cotal transmission, from 1937 to 1938. These motorcars were not only able to be operated remote, but were also equipped with a 150-horsepower diesel engine on the bogie, for convenient repair and replacement, and a streamlined nimble body with three tones of grayish Surrey green, clear yellow with a red tinge, and light blue. Nonetheless, the motor car equipped with an internal combustion engine was in "a transition period, with many models," and a specific model based on the SMR's own technology was not yet established. The development and introduction of motorcars equipped with internal combustion engines anticipated the introduction of a diesel locomotive.

A diesel locomotive was suitable for long distance and high-speed operation because it had good thermal efficiency and reduced the need for refueling; its driving operation was also easy, providing good acceleration. High-speed locomotives with an average speed of 130 km were manufactured for "Der Fliegende Hamburge" in Germany and for the "City of Denver" in the US. Globally, for high-speed trains with an average speed of more than 100 km, the diesel locomotive held the top position in the United States, Germany, and France beginning in 1937. To allow the introduction of diesel locomotives, though, the SMR purchased a 750-horsepower Dikii type of locomotive from Sulzer Ltd. of Switzerland and a 700-horsepower Dikini type locomotive from Mamse of Germany, respectively, in 1931, and produced a diesel-electric locomotive equipped with a 750-horsepower diesel engine domestically via Niigata Engineering Co., Ltd. Overall, however, the diesel engine was not fully introduced. Although the SMR had caught up with the technical state-of-the-art in steam locomotives, the introduction and localization of the diesel engine as a new technology were unsolved issues.[6]

As described above, even though the management of all Manchurian Railways by the SMR was an opportunity to expand the high technology of the SMR to all Manchurian Railways, and the Manchurian National Railways improved their technology level, the SMR faced a technical problem waiting for a solution.

4.5 Wartime Type of Technology and Its Post-war Historical Limits

The SMR tried to promote transportation capacity and enhance the completeness of the transportation network in Manchuria, in accordance with "the First Industrial Development 5-year Plan" in 1937. Nonetheless, because the Marco Polo Bridge Incident occurred on July 7 of the same year and the battle grew to become the second Sino-Japanese War, the SMR worked with railway corps of the Japanese army to take control of military transport and restore railway facilities damaged by the war. In August 1937, the North China Secretariat was established, operating the railways in North China. Even though its business scope was expanding to Northern China, initially the SMR did not manage to integrate the railways of North China into the existing system because the Japanese government and the Kwantung Army opposed such integration. Finally, in April 1939, the North China Railway Co. was established.

Likewise, the SMR's advancement towards North China railways and the establishment of a new railway company caused a significant outflow of its internal management resources. Various materials and railway vehicles had to be delivered, involving the staff of the SMR. The North China Railway Co. improved its railway operation based on the management resources of the SMR. The railway car repair method was introduced by the SMR to reduce the average number of days needed for repair in railway workshops, and the improvement of car dispatch capabilities and the standardization of yard work were developed so the North China Railway Co. could achieve a high labor productivity that was never experienced before the war (Lim 2006). This implies that the railway technology of the SMR was successfully applied to the northern part of China.

Under these circumstances, the SMR had to continue building railways not only to support its industrial development but also to fulfill the strategy of the Soviet Union. As shown in Fig. 4.6, railroad lines in operation totaled more than 11,000 km in the 1940s. Since railway facilities were exposed to severe cold in

[6] "A continent express, equipped with a 5,000-horsepower diesel engine, sleeping car, dining car, and observation car, which is capable of shortening the long distance of intercity transportation, for example between Busan and Shanghai and between Changchun and Nanjing, will speed up in the not-so-distant future" (Yoshino 1939, 34–35).

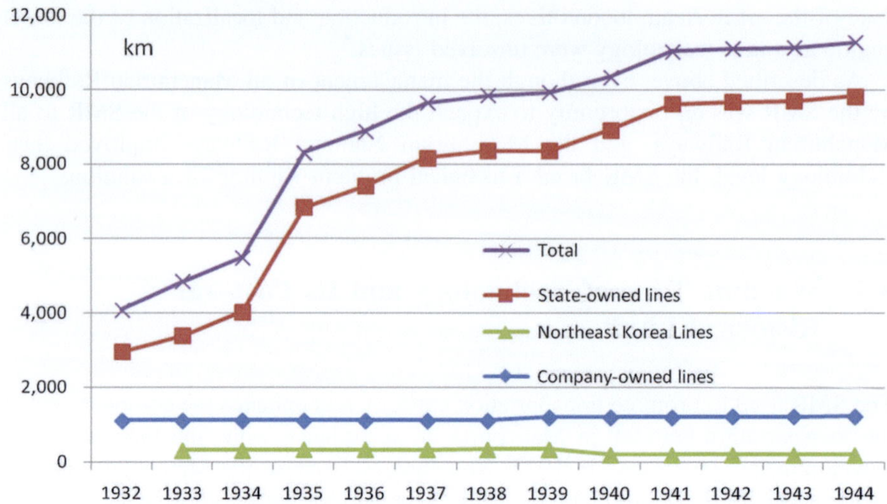

Fig. 4.6 Railroad route length of SMR after the manchurian incident. *Source* Minami Manshū Tetsudō (each month), Minami Manshū Tetsudō (each year), Keizai Chōsakai (1934), Keizai Chōsakai (1935), Minami Manshū Tetsudō (1938), Mantsukai (1986), Nihon Tōkei Kyōkai (1988), and Foreign Documents Division (1955)

these areas during the winter season, the Railway Technology Institute began promoting research on rail damage, frost on tracks and structures, protection against cold, and cold-weather concreting, following the decision of the Ad Hoc Committee for Intense Cold Measures, created in February 1937 (Mantetsu Tetsukenkai 1990, 61). At the same time, the institute continued to implement tests and conduct research on the use of substitutes, the cost reduction of construction and maintenance, and the strengthening of transportation capacity in response to the shortage of materials, and took advantage of the research results by improving tracks, communication, signal, engine shed, and station facilities.

Meanwhile, the SMR established a track-strengthening plan in 1939, to unify the "track maintenance knowledge" between company-owned and state-owned lines, and to maintain different kinds of tracks (Mantetsukai 1986, 267). Regarding construction standards, the SMR established a "Manchuria Railway Construction Revision Committee," with a Chief of Planning Department, and a Railway Construction Office as secretary-general, composed of experts from various institutions, such as construction, engineering, transportation, and driving, to conduct research and discuss these subjects after the inauguration of the Railway Secretariat General. However, the equipment standards were not decided until 1941 (Mantetsukai 1986, 249–254). On the other hand, the failure of track maintenance due to shortage of materials caused frequent operation accidents after the outbreak of the Second Sino-Japanese war (Chōsaka 1941). In addition, because land transportation started to cope with the shortage of maritime transport capacity after the outbreak of the Pacific War, the mainstream transportation route changed from the North-South direction of the Lianchang and Changbin line to the East-West direction of the Fengshan and Anfeng line. When the shortage of materials became significant, the SMR closed part of a double-track section of the Lianchang line and made it into a single-track section, diverting the rail to the completion of the double-track project of the Anfeng line. As the shortage of railway transport capacity worsened, the SMR inevitably stopped the operation of the Asia Express, "a symbol of the Manchurian culture."

On the other hand, railway vehicles nearly doubled in 1945 compared to 1936, especially vehicles for freight transportation. The SMR deployed 2,068 locomotives in the company-owned lines and state-owned lines: 308 locomotives for passenger transportation, 1,525 locomotives for freight transportation, 56 locomotives for the Railway Construction Office, 272 locomotives lent to other railways, and 29 locomotives for other purposes in 1945 (Ichihara et al. 1972, 108). Above all, a reduction of the vehicle models in the State Line was promoted, and most new locomotives were standard locomotives, such as the Paciro type of 2C1 high-pressure steam tender locomotive and the Mikai or Mikaro type of 1D1 high-pressure tender locomotive. "Especially, the Mikaha type of 1D1 high-pressure steam tender locomotive, made in Czechoslovakia, and the Sorina type of 1D high-pressure steam tender locomotive, made in the US, were purchased, the Paciha type of 2C1 high-pressure steam tender locomotives was introduced, and the 1D1 high-pressure steam condensing tender locomotive was also introduced as a trial." (Ichihara et al. 1972, 305) (Table 4.2).

As tension with the Soviet Union grew in 1941, the SMR manufactured the Mikaku type of steam locomotive, equipped with a condenser that collected water and that had been used once to enable a non-water supply run of 1600 km without

Table 4.2 Railroad car of SMR (unit: car)

	Locomotive	Passenger car	Freight car	Motorcar	Hand truck
1934	916	1,265	15,179	83	70
1935	1,094	1,525	18,229	101	106
1936	1,250	1,701	20,021	112	119
1937	1,384	2,109	21,974	120	176
1938	1,497	2,239	23,657	124	176
1939	1,617	2,404	26,987	123	178
1940	1,786	2,609	30,056	133	177
1941	1,946	2,814	32,353	141	174
1942	2,100	2,984	35,781	141	173
1943	2,265	3,194	39,309	141	171
1944	2,399	3,049	41,826	141	169
1945	2,422	3,061	41,984	141	171

Source Foreign Documents Division (1955, 239)

Note Though most of the statistics are based on the end of fiscal year, the data of 1945 are based on July 1945

water supply, for military purposes in Siberia, where it was difficult to secure water for the engine (Ichihara et al. 1972, 310–312). The SMR designed the Rikui type and the Rikuni type of 1D2 high-pressure steam tender locomotives, equipped with a booster and a two-axle bogie as a trailing truck, to improve the performance of the Mikai type and the Mikaro type of 1D1 high-pressure steam tender locomotive, which became the standard freight locomotive (Ichihara et al. 1972, 316).

When developing these new vehicles, the Railway Technology Institute carried out testing and research on the assumption that railway workshops would produce them, and the new vehicles were designed by the Machinery Section or the Railway Technology Institute itself. In particular, the enhancement of transport capacity, measures of inferior coal, the performance improvement of existing facilities, the accumulation of basic data on vehicle design and operation, the lightening of the vehicles, and the self-production of measuring instruments and testing machines were among the topics most commonly researched by the Railway Technology Institute, and the results were implemented in vehicle production and operation (Mantetsu Tetsukenkai 1990, 61). However, as the shortage of materials became a bottleneck in vehicle production, the self-development of diesel locomotives was no longer possible, creating a new technical problem, as described above. Meanwhile, as buying railway vehicles from outside was also difficult, the SMR focused on self-produced railway vehicles, with the Dalian workshop as a center of production. As shown in Fig. 4.7, after locomotive production in the Dalian workshop was significantly reduced by the Showa Depression, it started to rapidly increase after the outbreak of the Sino-Japanese war, producing about 30–40 locomotives in the 1940s.

The pivotal workshop of the SMR was the Dalian workshop, whose workload slightly decreased in wartime, but as the Mudanjiang workshop was established, the share of the workload of the State Line increased. The workload of railway

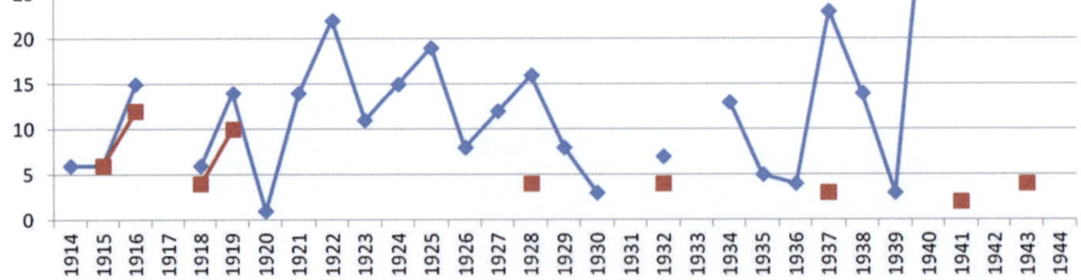

Fig. 4.7 Steam locomotive manufactured by Shahekou Railway Workshop, SMR (unit: car). *Source* Ichihara et al. (1972, 504–505)

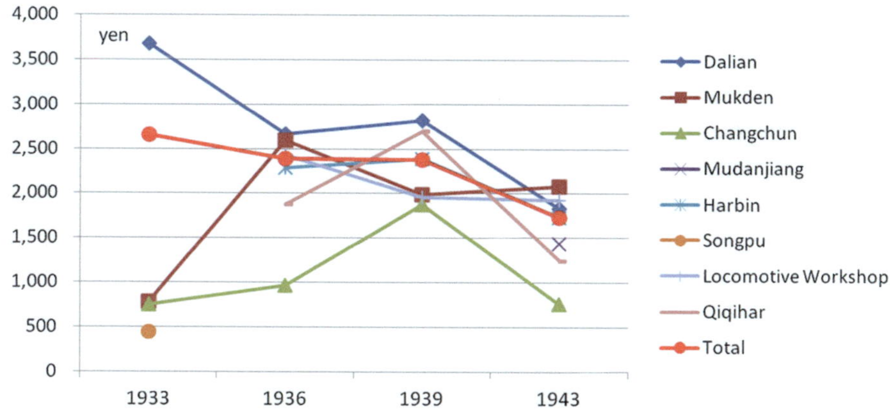

Fig. 4.8 Labor productivity of railway workshop, SMR (unit: yen). *Source* Minami Manshū Tetsudō (each year), and Minami Manshū Tetsudō (1943). *Note* Labor productivity 3 of Fig. 4.5 shown above

workshops was divided into: (i) new production; (ii) remodeling; (iii) repair; and (iv) other work. The Dalian workshop was mainly in charge of new production, while the workshops of state-owned lines were mainly responsible for repair and other work. In this study, we converted the work of all railway workshops and divided it by the number of workshop employees to estimate the trend of labor productivity, and we found that overall labor productivity decreased; this trend was remarkable in the 1940s, when the workload gap between the Dalian workshop and other workshops of state-owned lines narrowed (Fig. 4.8). By the end of the 1930s, the decreasing trend was remarkable, and, in 1943, the labor productivity of the Mukden workshop exceeded that of the Dalian workshop. However, the technological level of state-owned lines' workshops was not necessarily close to that of the Dalian workshop because the types of work performed were largely different, even though the technological level of state-owned lines certainly improved after railway unification (Table 4.3).

How can the overall decline in labor productivity in the 1940s be explained? In terms of days needed for repair in railway workshops, the waiting time for repair of locomotives and freight cars tended to increase during the war.[7] The labor composition saw an increase in non-skilled young workers, while workshops lacked

[7]The changes in the number of days for repair in railway workshops (per car) from 1938 to 1942 were 18.5 days → 19.9 days → 19.7 days → 20.4 days → 20.6 days for a locomotive; 17.8 days → 17.2 days → 21.3 days → 15.3 days → 16.5 days for a coach; and 7.6 days 9.4 days → 10.1 → 12.1 days → 13.6 days for a wagon (Tōkeisho 1990).

Table 4.3 Work composition and share of railway workshop, SMR (unit: %)

		Work composition of workshop			Work weight of workshop		
		1936	1939	1943	1936	1939	1943
Total	Subtotal	100.0	100.0	100.0	100.0	100.0	100.0
	Production	19.4	24.4	15.9	100.0	100.0	100.0
	Remodeling	13.2	4.6	1.4	100.0	100.0	100.0
	Maintenance	38.9	42.7	38.4	100.0	100.0	100.0
	Others	28.5	28.3	44.3	100.0	100.0	100.0
Dalian	Subtotal	100.0	100.0	100.0	52.5	45.0	44.2
	Production	37.0	45.5	35.6	100.0	84.0	99.3
	Remodeling	4.7	3.6	1.5	18.7	35.4	45.3
	Maintenance	31.8	26.1	21.5	42.9	27.5	24.8
	Others	26.5	24.8	41.4	48.8	39.4	41.3
Mukden	Subtotal	100.0	100.0	100.0	20.1	19.5	21.5
	Production	0.0	0.0	0.5	0.0	0.0	0.6
	Remodeling	24.5	13.6	2.1	37.3	57.2	31.3
	Maintenance	52.4	61.9	48.9	27.1	28.2	27.4
	Others	23.0	24.4	48.6	16.3	16.8	23.6
Changchun	Subtotal	100.0	100.0	100.0	3.6	6.5	2.9
	Production	0.0	4.4	0.0	0.0	1.2	0.0
	Remodeling	16.2	0.9	0.0	4.4	1.3	0.0
	Maintenance	12.4	66.7	63.7	1.1	10.1	4.9
	Others	71.4	28.0	36.3	9.1	6.4	2.4
Mudanjiang	Subtotal			100.0			4.6
	Production			0.0			0.0
	Remodeling			0.0			0.0
	Maintenance			55.5			6.7
	Others			44.5			4.7
Harbin	Subtotal	100.0	100.0	100.0	15.4	15.4	11.7
	Production	0.0	23.4	0.1	0.0	14.8	0.0
	Remodeling	32.1	0.5	1.7	37.4	1.7	13.9
	Maintenance	39.6	40.4	47.8	15.7	14.6	14.6
	Others	28.3	35.7	50.5	15.3	19.4	13.3
Locomotive workshop	Subtotal	100.0	100.0	100.0	5.0	9.0	8.8
	Production	0.0	0.0	0.0	0.0	0.0	0.0
	Remodeling	1.9	1.4	1.0	0.7	2.6	6.1
	Maintenance	62.9	61.6	52.0	8.1	13.0	11.9
	Others	35.2	37.0	47.0	6.2	11.8	9.3
Qiqihar	Subtotal	100.0	100.0	100.0	3.5	4.6	6.2
	Production	0.0	0.0	0.0	0.0	0.0	0.0
	Remodeling	5.5	1.8	0.8	1.5	1.8	3.5
	Maintenance	57.7	60.1	60.5	5.1	6.5	9.8
	Others	36.7	38.1	38.7	4.5	6.2	5.4

Source Minami Manshū Tetsudō (each year), and Minami Manshū Tetsudō (1943)

Table 4.4 Railway workshop engineer training school (unit: person)

Railway workshop			Japanese	Chinese	Total
Dalian	Regular course	Major part	421		421
		Minor part	31		31
	Regular course	Major part			
		Minor part			
	Subtotal		452		452
Mukden	Regular course	Major part	101		101
	Regular course	Major part			
		Minor part		94	94
	Subtotal		101	94	195
Changchun	Regular course	Major part	43		43
	Regular course	Major part			
		Minor part		35	35
	Subtotal		43	35	78
Harbin	Regular course	Major part	145		145
	Regular course	Major part			
		Minor part		124	124
	Subtotal		145	124	269
Qiqihar	Regular course	Minor part		39	39
Total			741	292	1,033

Source Minami Manshū Tetsudō (1989)

engineers (Takahashi 1940, 43–52). Although the functional foreman system of Taylor was being adopted at that time, there was a qualitative deterioration of the labor force and, together with the lack of materials, this caused a decrease in the efficiency of work. Therefore, a re-education of workers was implemented inside railway workshops.

In addition to the Railway Engineer Training School inside the SMR, other Railway Workshop Engineer Training Schools were created as engineer training institutions to teach workers to perform tasks including assembly, finishing, lathe work, internal combustion engine work, riveting, modeling, iron work, brass work, tool work, welding, electrical work, and gas work, as shown in Table 4.4. For the purpose of training an engineer, as soon as a graduate of higher elementary school or anyone under 18 years of age entered the school, after an entrance test, the individual was hired as an employee of the SMR and began living in the dormitories of the educational institutions. A graduate was given a status of lower employee (Yōnin) or higher employee (Koin, only for the Japanese after 1942), and an occupational title of repair worker or mechanic worker. They were qualified to be promoted to a higher employee or assistant status (Junshokuin, only for the Japanese after 1942) through a screening process three years after graduation.

A regular course was conducted for only the Japanese, and a special course was offered to a minority of the Chinese. This study emphasizes that, since the labor system in workshops was characterized by an ethnic hierarchy, education also had the function of reproducing ethnic disparities. Additionally, the Railway Trainee System was provided only to the Chinese, who numbered 199 in Mukden, 83 in Harbin, 49 in Sankeshu, and 48 in Qiqihar, for a total of 299 trainees in all railway workshops (Table 4.5).

Thus, a Japanese employee was considered a higher engineer, while a Chinese employee considered a lower engineer, and education opportunities were offered only to some of the Chinese and only within the internal education system of railway workshops. These institutions were mainly introduced to enhance the technical level of the Chinese who were newly transferred to the SMR after the Manchurian Incident, and the SMR singlehandedly trained engineers who became difficult to employ in the external labor market and gave an opportunity of promotion to graduates. The internal labor market expanded through the training system during the war. As for ethnic status in workshops, the Japanese were promoted to a higher status, such as high-level staff (Shokuin) and assistant staff (Junshokuin), in the company-owned lines or state-owned lines during the war, and the status of lower employees (Yōnin) for the Japanese disappeared. In 1942, the lowest possible status for a Japanese employee was that of a higher employee (Koin). In contrast, the Chinese showed a tendency to be concentrated in the lower status. The Chinese continuously replaced the Japanese in the higher positions, and the Chinese became a majority as higher employees (Koin) and mid-level status employees. This trend was more significant in state-owned lines, where the Chinese were the majority from the beginning, than in the Company Line.

The ethnic skill gap was still not resolved, not even during the war. For example, according to the results of the skill survey conducted on 854 workers, 503 Japanese and 351 Chinese employees who served in the lathe workplace, finishing workplace, and bogie workplace of the Dalian railway workshop in 1942, the skill composition of the Japanese was A 13.9%, B 33.9%, C 29.0%, D 19.4%, and E 3.8%, while that of the Chinese was A 12.3%, B 26.6%, C 31.8%, D 18.9%, and E 10.3%. Even though skill composition had a close relationship with the number of years of service, Class-D Chinese lathe workers had 23.4 years of service, suggesting that other factors, such as education and work content, greatly affected this classification (Hayashi 1942). Such skill disparities were reflected in duties and positions in railway workshops, and higher duties were primarily limited to the Japanese.

Likewise, the lack of materials and labor force became significant during the war, and the wartime type of technology was developed to cope with it. Nonetheless, the technical hierarchy based on ethnic groups remained strongly attached to the colonial employment structure.

Table 4.5 Status composing of ethnic groups in railway workshop of SMR (unit: person)

		Company lines			State lines			Total		
		Japanese	Chinese	Subtotal	Japanese	Chinese	Subtotal	Japanese	Chinese	Subtotal
High-level staff (Shokuin)	1935.3	187		187	152	64	216	339	64	403
	1938.3	189	1	190	229	63	292	418	64	482
	1940.3	287	1	288	292	62	354	579	63	642
	1943.3	437		437	495	56	551	932	56	988
	1944.3	625	6	631	682	66	748	1,307	72	1,379
	1944.9	567	6	573	797	87	884	1,364	93	1,457
Assistant staff (Junshokuin)	1935.3									
	1938.3									
	1940.3									
	1943.3	781	23	804	576	76	652	1,357	99	1,456
	1944.3	912	149	1,061	670	152	822	1,582	301	1,883
	1944.9	782	140	922	831	355	1,186	1,613	495	2,108
Higher employees (Koin)	1935.3	262		262	157	58	215	419	58	477
	1938.3	403	7	410	231	73	304	634	80	714
	1940.3	618	10	628	310	88	398	928	98	1,026
	1943.3	2,126	599	2,725	2,463	1,232	3,695	4,589	1,831	6,420
	1944.3	1,716	516	2,232	2,376	1,444	3,820	4,092	1,960	6,052
	1944.9	1,053	433	1,486	1,566	1,603	3,169	2,619	2,036	4,655
Lower employees (Yōnin)	1935.3	1,190	928	2,118	5	2,198	2,203	1,195	3,126	4,321
	1938.3	1,986	1,404	3,390	954	3,304	4,258	2,940	4,708	7,648
	1940.3	2,722	1,877	4,599	1,882	5,216	7,098	4,604	7,093	11,697
	1943.3		2,850	2,850	–	5,499	5,499		8,349	8,349
	1944.3	–	5,336	5,336	–	6,252	6,252	–	11,588	11,588
	1944.9		3,347	3,347	–	6,411	6,411		9,758	9,758

(continued)

Table 4.5 (continued)

		Company lines			State lines			Total		
		Japanese	Chinese	Subtotal	Japanese	Chinese	Subtotal	Japanese	Chinese	Subtotal
Non-regular employee (Shaingai)	1935.3	926	1,292	2,218	27	298	325	953	1,590	2,543
	1938.3	200	749	949	102	1,613	1,715	302	2,362	2,664
	1940.3	38	655	693	63	2,384	2,447	101	3,039	3,140
	1943.3	169	10	179	89	1,437	1,526	258	1,447	1,705
	1944.3	172	6	178	91	1,468	1,559	263	1,474	1,737
	1944.9	114		114	75	778	853	189	778	967
Total	1935.3	2,565	2,220	4,785	341	2,618	2,959	2,906	4,838	7,744
	1938.3	2,778	2,161	4,939	1,516	5,053	6,569	4,294	7,214	11,508
	1940.3	3,665	2,543	6,208	2,547	7,750	10,297	6,212	10,293	16,505
	1943.3	3,513	3,482	6,995	3,623	8,300	11,923	7,136	11,782	18,918
	1944.3	3,425	6,013	9,438	3,819	9,382	13,201	7,244	15,395	22,639
	1944.9	2,516	3,926	6,442	3,269	9,234	12,503	5,785	13,160	18,945

Source Minami Manshū Tetsudō (each year), Minami Manshū Tetsudō (1938, 1113–1133), Minami Manshū Tetsudō (1943), and Minami Manshū Tetsudō (1989)

4.6 Conclusion

The SMR supported the policies of the Japanese Government, as a Japanese version of the "East India Company." However, Japanese technology did not secure a standard-gauge railroad and the SMR had to import railroad vehicles and materials from the US after standard-gauged reconstruction work began. The SMR temporarily purchased locomotives made in the UK, but the American type of locomotive was finally chosen because of the simplicity of the structure and maintenance of its railroad cars. In the process, foreign technology was absorbed by the SMR through the ordinary management and localization of railway technology. In addition, the reinforcement of tracks using heavy rails and the introduction of heavy locomotives allowed the SMR to introduce longer and faster trains.

In the railroad workshop, the SMR not only implemented the expansion of existing facilities through workplace enlargement and the installation of various machines but also enhanced work efficiency through the introduction of internal educational institutions and a Rowan premium wage plan. Initially, railway workshops accomplished the domestic production of passenger and freight cars to meet the challenge of producing a locomotive with a high level of technology. The Solisi type of freight engine was produced as a dead copy in 1914, and domestic engine production continued afterward. The Pasiko type of engine, equipped with a combustion chamber in one engine, was produced experimentally, following the SMR's original design, in 1927. In terms of passenger and freight cars, technological progress, such as the introduction of the steel frame, increase in solidity, dead weight reduction of cars, and design simplification, was significant.

This railroad car production, following the SMR's original design, was achieved through the synergies realized among the research of the Railway Technology Institute, design technique of the Machinery Section, and manufacturing technique of the Railway Workshop. Engineers were employed in Manchuria and Japan, and skilled engineers in charge of the vehicle development were then sent to a research institute, a workshop, or the machine section. A routing system and Taylorism were introduced into the railway workshop and a significant repair time reduction and labor productivity improvement were achieved. The railway technology of the SMR was adapted to Mainland China in a sophisticated way.

When the Manchurian Incident happened, the railway technology of the SMR spilled over to all Manchurian railways. A Facilities Committee was established in the Railway General Office. The unification of railroad tracks and cars was accomplished, and, eventually, the company-owned lines, state-owned lines, and Northeast Korea lines were unified in 1936. The workshop management technology accumulated in company-owned lines spread through the workshops of state-owned lines, the transfer of Japanese engineers, regular workshop manager meetings, workshop study groups, business meetings, and the rationalization of railway workshops. In the meantime, the innovation of railroad car production continued, and the Asia Express train was the climax of the SMR's technology.

With the outbreak of the Second Sino-Japanese war, even though buying material from the outside became difficult, a large amount of managerial resources from the SMR flowed toward Northern China, and the construction of new lines including lines from Manchukuo to the Soviet border, continued. Thus, the lack of materials became a serious issue. However, the demand for transport increased consistently, and wartime standards were implemented in both the maintenance of railroad facilities and railroad car production. Internal engineer education and self-produced railroad cars were supported. Despite such measures, driving accidents, due to the defective maintenance of tracks, and repairs occurred frequently. A basic maintenance of the railroad, through the conversion of existing facilities, was carried out, and the Asia Express service stopped operating.

As described above, the SMR's technical accumulation became world-class in the prewar period, and this level of development was also present in the organization's internal management. In addition, the most advanced technology was realized based on superior management conditions. The wartime type of technology was developed in response to the lack of resources experienced during the war, but it was not able to solve new technical problems, such as dieselization. The core technology was confined to Japanese employees and technology spillovers to Chinese employees were restricted. Therefore, the stay of Japanese engineers and the support of the Soviet Union engineers were required for the postwar management of the Chinese railroads.

References

Amano, Hiroyuki. 2012. *Mantetsu tokkyū ajia no tanjō (Birth of SMR Asia Express)*. Tokyo: Harashobō.

Asō, gorō, 1933. *Saishin tetsudō kōjō (Newest railway workshop)*, Tokyo: Tetsudō Jihō kyoku.

Bae, Bugil. 2002. *Manshūkoku to keieigaku (Manchukuo and Business Administration)*. Tokyo: Nihon Tosho Sentā.

Chōsaka, Chōsakyoku, Minami Manshū Tetsudō Kabushikigaisha (Investigation Department, Investigation Section, SMR). 1941. *Kaisha tetsudō gyōmu no gyōmu chōsa daiichiji chūkan hōkoku (Business research on railway business of company the first interim report)*.

Chūgai Sangyō Chōsakai. (ed.). 1923. *Nōritsu zōdai kagaku teki shimatsuhō hōkokusho dai san satsu: Chinuke shigoto no soshiki oyobi kantokuhō (Efficiency increase science teport vol. 3: Method of organization of subcontract work and supervision)*, Tokyo: Chūgai Sangyō Chōsakai.

Foreign Documents Division, CIA, USA. 1955. *Transportation in northeast China*, Washington D.C.: Foreign Documents Division Translation, No. 354.

Hayashi, Shigeki, Minami Manshū Tetsudō Kabushikigaisha Shinkyō Shisha Chōsashitsu (Investigation Office, Changchun Branch, SMR). 1942. *Rōmu kanri chōsa chūkan kōkoku sono ichi: Dairen tetsudō kōjō ni okeru manjin kōsakukō no zenreki jittai chōsa (Labor management investigation interim report 1: Career investigation of Chinese workers in Dalian workshop)*.

Ichihara, Yoshidzumi, Yoneo Oguma, Ryūsaburō Nagata, and An'yōji Osamu (eds.). 1971. *Minami mansu tetsudō: ajiagō to kyakukasha no subete (SMR: Asia Express and passenger, freight car)*. Tokyo: Seibundo-shinkōsha.

Ichihara, Yoshidzumi, Yoneo Oguma, Ryūsaburō Nagata, and An'yōji Osamu (eds.). 1972. *Minami mansu tetsudō: Tetsudō no hatten to kikansha (SMR: Develop of railway and locomotive)*. Tokyo: Seibundo-shinkōsha.

Ichihara, Yoshidzumi. 2010. *Mantetsu tokkyū ajiagō (Asia Express of SMR)*. Tokyo harashobō.

Kaise, Kingo. 1931. Nōritsu gakari tanjō ki (Record on birth of efficiency subsection). *Kyōwa (Harmony)*. 44: 4–5.

Keizai Chōsakai, Minami Manshū Tetsudō Kabushikigaisha (SMR), (ed.). 1934. *Manshū keizai tōkei zuhyō (Manchurian economy statistical chart and table)*. Dailen: Minami Manshū Tetsudō Kabushikigaisha (SMR).

Keizai Chōsakai, Minami Manshū Tetsudō Kabushikigaisha (SMR), (ed.). 1935. *Manshū kōtsū tōkei shūsei (Manchurian transport statistics collection)*. Dailen: Minami Manshū Tetsudō Kabushikigaisha (SMR).

Lim, Chaisung. 2006. Nitchū sensō-ka no kahoku kōtsū no setsuritsu to senji yusō no tenkai (The establishment of the North China Railway Company and wartime transportation during the Sino-Japanese war, 1937–1945). *Rekishi to keizai (The journal of political economy and economic history)*. 49–1: 1–15.

Lim, Chaisung. 2010. Senzen-ki kokutetsu niokeru tetsudō un'ei kanri no tokushitsu to naibu gōrika (A characteristics of the railroad management and internal rationalization of JNR in prewar period) In *RyōTaisenkanki no toshi kōtsū to un'yu (Japanese urban transportation during the interwar period)*, eds., Yoshinobu Oikawa. Tokyo: Nihonkeizaihyōronsha.

Lim, Chaisung. 2013. Mantetsu ni okeru tetsudōgyō no tenkai: Kōritsusei to shūekisei no shiten yori (The South Manchuria Railway Company's development of the railroad business: An examination of efficiency and profitability). *Rekishi to keizai (The journal of political economy and economic history)*. 55–4: 1–15.

Mantetsu Shainkai (SMR Employee Society). 1929. Mantetsu gyōtai 1: shakakō kōjō (SMR business condition 1: Shahekō workshop). *Kyōwa (Harmony)*. 50: 10–12.

Mantetsu Shainkai (SMR Employee Society). 1936a. Kyōwa Mantetsu shinbun (Harmony newspaper of SMR). *Kyōwa (Harmony)*. 177: 24.

Mantetsu Shainkai (SMR Employee Society). 1936b. Sōkyoku an'nai (Railway secretariat general guidance). *Kyōwa (Harmony)*. 180: 18–19.

Mantetsu Shainkai (SMR Employee Society). 1937. Kyōwa Mantetsu shinbun (Harmony newspaper of SMR). *Kyōwa (Harmony)*. 177: 26.

Mantetsu Tetsukenkai. 1990. *Mantetsu tetsudō gijutsu kenkyūjoshi (History of SMR Railway Technology Institute)*. Kumamoto: Mantetsu Tetsukenkai.

Mantetsukai (SMR society), (ed.). 1986. *Minami manshū tetsudō kabushikigaisha daiyoji jyūnenshi (The forth 10 years outline history of SMR)*. Dailen: Ryūkeishosha.

Minami Manshū Tetsudō Kabushikigaisha (SMR), (ed.). 1919. *Minami manshū tetsudō kabushikigaisha jyūnenshi (The 10 years outline history of SMR)*. Dailen: Minami Manshū Tetsudō Kabushikigaisha (SMR).

Minami Manshū Tetsudō Kabushikigaisha (SMR), (ed.). 1928. *Minami manshū tetsudō kabushikigaisha dainiji jyūnenshi (The second 10 years outline history of SMR)*. Dailen: Minami Manshū Tetsudō Kabushikigaisha (SMR).

Minami Manshū Tetsudō Kabushikigaisha (SMR), (ed.). 1937. *Minami manshū tetsudō kabushikigaisha sanjyūnen ryakushi (The 30 years outline history of SMR)*. Dailen: Minami Manshū Tetsudō Kabushikigaisha (SMR).

Minami Manshū Tetsudō Kabushikigaisha (SMR), (ed.). 1938. *Minami manshū tetsudō kabushikigaisha daisanji jyūnenshi (The Third 10 years outline history of SMR)*. Dailen: Minami Manshū Tetsudō Kabushikigaisha (SMR).

Minami Manshū Tetsudō Kabushikigaisha (SMR), (ed.). 1943. *Tetsudō tōkei nenpō (Annual railway statistics report)*. Dailen: Minami Manshū Tetsudō Kabushikigaisha (SMR).

Minami Manshū Tetsudō Kabushikigaisha (SMR), (ed.). each month. *Tōkei geppō (Monthly statistics report)*. Dailen: Minami Manshū Tetsudō Kabushikigaisha (SMR).

Minami Manshū Tetsudō Kabushikigaisha (SMR), (ed.). each year. *Tōkei nenpō (Annual satistics report)*. Dailen: Minami Manshū Tetsudō Kabushikigaisha (SMR).

Minami Manshū Tetsudō Kabushikigaisha (SMR). 1929. *Shakō kōjō rōdō sōgi gaiyō (Labor dispute summary in Shahekou workshop)*. Minami Manshū Tetsudō Kabushikigaisha (SMR): Dailen.

Minami Manshū Tetsudō Kabushikigaisha. 1989. *Zaiseki shain tōke: Shōwa jyūkyūnen kyūgatsumatsu genzai (Enrolled employee statistics: The end of September 1944)*. Tokyo: Ryūkeishosha.

Nagami, Toshiaki. 2003. Mantetsu no tetsudō gijutsu iten to chūgoku no tetsudō fukkō (The transfer of South Manchuria railway technology in China after world war 2). *Nihon shokuminchi kenkyū (The journal of Japanese colonial studies)*. 15: 1–17.

Nihon Tōkei Kyōkai (Japan Statistical Association). 1988. *Nihon chōki tōkei sōran (Historical statistics of Japan)*.

Ōkawa, Kazushi, Tsutomu Noda, Nobukiyo Takamatsu, Yamada Saburō, Minoru Kumazaki, Shionoya Yūichi, and Minami Ryoshin. 1966. *Bukka (Prices)*. Tokyo: Tōyōkeizaishinpōsha.

Ōnaka, Shinobu. 1940. Nishida shi to nōritsu (Mr. Nishida and efficiency), *Manshū no nōritsu (Manchurian efficiency)*. 2–1.

Rinji Keizai Chōsa Iin-kai, Minami Manshū Tetsudō Kabushikigaisha (Temporarily Economic Investigation Committee, SMR), (ed.). 1929. *Shakō kōjō ni okeru nikkaryōkō no nōryoku hikaku (Ability comparison between Japanese and Chinese workers in Shahekou workshop)*. Dailen: Minami Manshū Tetsudō Kabushikigaisha (SMR).

Sato, Yoshihiko. 2002. Mantetsu niokeru ajiagō no unten to tetsudō gijutsu kenkyūjo ga hatashita yakuwari (Operation of Asia Express and role of railway technology institute in SMR). *Tetsudō shigaku (Japan railway history review)*. 20: 33–42.

Sawai, Minoru. 1998. *Nihon tetsudō sharyō kōgyōshi (History of Japanese railway vehicle)*. Tokyo: Hihon Keizai Hyōronsha.

Senkōkai. (ed.), 1986. *Chōsen kōtsū shi (Korea transportation history)*. Tokyo: Senkōka.

Shigeru, Tomekichi. 1936. Ichigenka made no tetsudōbu (Railway section until unification). *Kyōwa (Harmony)*. 179: 20–21.

Shiryōka, Chōsa-bu, Minami Manshū Tetsudō Kabushikigaisha (Document Section, Investigation Department, SMR), (ed.). 1942. *Tetsudō tōkei tekiyō (Railway statistics summary). Vol. 8*. Dailen: Minami Manshū Tetsudō Kabushikigaisha (SMR). November 10, 1942.

Shomuka, Tetsudōbu, Minami Manshū Tetsudō Kabushikigaisha (General Affairs Department, Railway Section, SMR), (ed.). 1930. *Kōsaku tōkei nenpō (Annual machinery work report)*. Dailen: Minami Manshū Tetsudō Kabushikigaisha (SMR).

Takahashi, Tadayuki. 1931. Kōjō kanri (Factory management). *Kyōwa (Harmony)*. 43: 16–17.

Takahashi, Tadayuki. 1940. Kōjō kanri jō no ichi kōsatsu (One consideration on the workshop Management). *Manshū no nōritsu (Manchurian efficiency)*. 2–1: 43–52.

Takahashi, Yasutaka. 1995. *Nippon shokuminchi tetsudō shiron (A historical study on Japanese colony railways)*. Tokyo: Nihonkeizaihyōronsha.

Tetsudō Sōkyoku, Minami Manshū Tetsudō Kabushikigaisha (Railway Secretariat General, SMR), (ed.). 1938. *Yōroppa sho tetsudō to hikaku seru Mantetsu shasen tetsudō shōsen no eigyō tōkei ichiran-hyō (The business statistics table of the company-owned lines and state-owned lines of SMR with comparison with the European railways)*.

Tetsudōbu, Minami Manshū Tetsudō Kabushikigaisha (Railway Section, SMR). 1932. *Tetsudō kōjō keiei hōshin narabini sharyō shūzen taisaku (Railway workshop management policy and railway vehicle repair measures)*.

Tōkeisho, Tōhoku Zaikei Iin-kai (Statistics Investigation Office, Northeast Financial Committee), (ed.). 1990. Kyū Manshū Keizai Tōkei Shiryō: "Wei man jiki tōhoku keizai tōkei" 1931–1945-nen (Former Manchurian economic statistics document: "Northeast economic statistics during the Manchukuo period" 1931–1945), Tokyo: Kashiwashobō.

Ueda, Kenji. 1935. Mantetsu ryūsenkeigata tokkyū ajiagō no unten to tetsudō keiei gijutsu jō ni okeru supīdo appu no mondai (Operation of streamline type Asia Express of SMR and problem of the speed up in the railway management technology) In *Manshūkoku keizai kenkyū (Manchukuo eonomy research*). Tokyo: Keiōgijukudaigaku Tōajijōkenkyūkai (Eastern Asia Affair Research Association, Keio Univesity).

Yoshino, Nobutarō. 1939. Kisha hattatsushi: Ameigata kara Ajia made (History of train development: from Amei type to Asia Express). *Kyōwa (Harmony)*. 251: 34–35.

Chapter 5
Innovation in Power Sources for Taiwan's Railways in the Period of US Aid (1950–1965)

Lung-Pao Tsai

Abstract It's worthy of consideration that the relationship between Taiwan, the US, and Japan in this period is formed of cooperation, dependence, and the issues formed by Taiwan obtaining foreign capital and suffering invasion of its capital. If a country wants to develop itself but lacks sufficient funds and technology, it is necessary to cooperate with or depend on other countries. Since the TRA did not attach importance to technical research and development, and expected the "Mainland China Recovery" to take place, it naturally cooperated with the US and Japan (which were the anti-communist alliance) to complete its technological innovations. Taiwan's development conditions were different from the United States' as well as Japan's. The partial dieselization policy adopted by the TRA had its own unique features and necessities, and met Taiwan's phased demands adequately at the time. In this process, the US well exceeded Japan in influence over Taiwan. Also, the US aid operations were highly beneficial for the US firms involved. The cooperation structure of Taiwan, the US, and Japan can be observed clearly in the TRA's development in the US-aid period. This structure's influence can clearly be seen in the long-term development of the TRA.

Keywords TRA · US-aid · Dieselization · Locomotive · Consulting company

5.1 Introduction

In the post–war period, Taiwan was successfully separated from colonial Japanese interests. The normalization and further development of the Taiwan Railway Administration (TRA) were achieved by the efforts of the nationalist government, retained Japanese employees, Taiwanese employees, and US aid, which started

---

L.-P. Tsai (✉)
Department of History, National Taipei University, 151, University Rd., San Shia District, New Taipei City 237, Taiwan
e-mail: henrytsai8@hotmail.com

M. Sawai (ed.), *The Development of Railway Technology in East Asia in Comparative Perspective*, Studies in Economic History,
DOI 10.1007/978-981-10-4904-0_5

from 1950. During this process, it was noteworthy how the different cultural and technological influences on Taiwan's railway system would be integrated.

Would the development of Taiwan's railway systems be moved towards the Chinese style because of the nationalist Taiwanese government, the American style because of the US aid, or would it keep with the Japanese style that had been used in the colonial period before the war? Would there be any development in the various dimensions of the industry—the administrative organization, personnel structure, hardware, system of fees, management, the organization's business model, and related elements? These questions involved not only economic issues but also post–war Taiwanese politics and diplomacy. In the interest of settling the above concerns, it is worthwhile to study the trains, technical matters, and innovation in power sources during Taiwan's US aid period, as well as the nuanced meanings of politics, economy, and diplomacy unfolding in the background.

Power innovation requires funds and technology. Since the beginning of US aid to Taiwan in 1950, foreign funds became the most essential part of the railways' capital structure. It is worth noting that these incoming foreign funds were often accompanied by new technology, and that competition between the US and Japan for the Taiwanese railway market was quite involved. This research aims to investigate the TRA's development during Taiwan's US aid period—especially the guidelines for power innovation, the roles of the US and Japan within what happened, and whether Taiwan has truly decolonized in the wake of Japanese rule or was recolonized by the US as a result of US aid.

5.2 US-Aid Operation and TRA Development

The second Chinese civil war broke out after the end of World War II, and during this war, Taiwan's Railways suffered some light damage. However, the Railway Department (RD)'s original construction was not solid, and the RD did not practice regular maintenance during the war. As a result of this, it was challenging to restore the Taiwan railway in the complicated post–war environment, especially with regard to human resources and materials. One major aspect of the human resources issue was that only a few Taiwanese employees were trained in the operation and care of the RD's mid–level or high–level technology.

Because of the lack of trained workers, as Japanese railway workers began repatriating the Railway Administration Council was forced to adopt a policy of going out of its way to retain Japanese talent in order to speed up the restoration. In this way, in the early post–war period a fast restoration was made possible by the concerted efforts of the Nationalist Government's elite cadre of leaders at the top of the organization, retained Japanese management talent in the middle, and the skillful Taiwanese employees at the bottom. However, the whole business aspect of the railway system was itself still out of order. The physical normalization and progress of the TRA would be developed successfully later, during the US Aid period (Tsai 2007, 27–48).

Innovation requires funds and technology. Therefore, in the US Aid period, foreign funds became the most essential part of the railway industry's capital composition, and these foreign funds were often accompanied by new technology. In the US–aid period from 1950 to 1965, the TRA obtained US aid through the Project Agreement, Counterpart Fund, and Development Loan Fund. The TRA obtained US Aid in the form of loans from 1950 to 1960. In these 11 years, the TRA applied for 47 loans totaling $36,134,000 USD from the Development Loan Fund, Project Agreement and the forth particular account and $147,265,000 NTD from the Counterpart Fund in total (see also the Table 5.1) (Council for International Economic Corporation and Development, Executive Yuan 1964, 3–4). While the US was providing aid to Taiwan, American consulting companies and railway professional advisers were also lending their services to the TRA while advanced machinery purchases were taking place and so on—this affected the TRA profoundly.

5.2.1 Roles of American Consulting Companies and TRA American Advisors

Based on the negative experience that the US had assisting the R.O.C. in the later period of World War II (Miller1974, 1–480)[1] as well as the similar experiences of the United Nations Relief and Rehabilitation Administration (UNRRA) and the Chinese National Relief and Rehabilitation Administration (CRE) assisting Taiwan in the post-war period (Kerr 2014, 161–186),[2] the applications for US aid were more carefully checked than those of past aid to ensure the efficient use of the funds and materials. For example, after the US aid-supported construction of the Central Cross-Island Highway started, 40% of the engineering costs were claimed by the Ret-Ser Engineering Agency, Veterans Affairs Commission, and the Directorate General of Highways as profits or management fees. This was found and corrected

[1]Referred to Miller, Merle (1974). *Plain Speaking: An Oral Biography of Harry S. Truman*. Putnam Publishing Group, "The American President, Harry S. Truman, repeatedly stressed that the US aid to China suffered severe defalcation. The United States Congress and Federal Bureau of Investigation found the abnormal financial activity of Kung, Hsiang-Hsi and Soong, Tse-Ven families, and they were very dissatisfied. Once President Truman said during an interview with the American author Merle Miller, and chewed out the Kuomintang, saying that that all of them were thieves that had stolen 750 million USD from the 3.8 billion USD sent to Chiang Kai-Shek.".

[2]Referred to Kerr (2014), *Formosa Betrayed* (pp. 161–186). Taipei: Taiwan Association of University Professors, "The UNRRA team worked in Taiwan until December of 1947. At this time, goods intended for relief valued at about $25,000,000 USD were brought to Taiwan (this valuation of the relief articles was made before the goods arrived in Taiwan). The UNRRA oversaw the dispatch and sell of these articles. People often paid high prices to buy these relief packages. These goods created profits many times over for the Bank of China, which was governed by the Executive Yuan under Tse-Ven Soong".

Table 5.1 Annual US-aid loan list

Year	Item	
	USD	NTD
1950	658,000	
1951	2,545,000	
1952	407,000	4,500,000
1953	1,283,000	11,000,000
1954	258,000	16,900,000
1955	429,000	1,375,000
1956	100,000	30,940,000
1957	5,264,000	39,061,000
1958	4,478,000	5,025,000
1959	10,155,000	38,464,000
1960	10,557,000	
Total	36,134,000	147,265,000

Source Council for International Economic Cooperation and Development, Executive Yuan (1964), p 5

by Valery S. de Beausset, the manager of the J. G. White Corporation, rescuing 100 million NTD from this fate (Council for US Aid, 1956, File ID: 064000000964A).

Regardless of past experiences with graft, the US provided an average of 100 million USD in economic aid per year for 15 years starting from 1950. Taiwanese economic development plans were the basis for applications to the US for economic aid, and were used to justify the ways in which the aid received would be used (Takeharu 1992, 17–18). In the examination and implementation processes, the American advisors and US consulting companies had an overall positive effect on Taiwanese economic development. However, there were still some disadvantages. For example, the application process took a lot of time, the amounts of annual aid that would be provided were uncertain and sometimes much less than the amount applied for, the policy about providing aid might change, the US might not respect the actual requirements or needs of the case or the planned policies and programs of R.O.C., and so on.

In the 1950s, most foreign exchange Taiwan required for construction depended on the US aid. To review the aid program, Taiwan and the US hired the J.G. White Corporation as engineering consultants. The J.G. White Corporation came to Taiwan in spring of 1949, and the manager on the case was Valery Sergei de Beausset. The firm helped plan many essential industrial programs and policies. The J.G. White Corporation's draft of industrial plans for the fiscal years 1952–1955 became the primary part of the first phase of plans by Taiwan's Council for Economic Planning and Development (CEPD). In addition, Richard C. Morse was hired as the TRA's adviser, and the organization was renamed 'the Council for United States Aid Engineering Consulting Group' in 1962.

The contract between J.G. White Corporation and the Council for United States Aid (CUSA) expired on July 31, 1962, but for the future development of Taiwan's industrial construction, CUSA needed foreign engineers to continue assisting with technical issues. Therefore, the CUSA signed a technical cooperation contract with the White Company for six engineers who had performed well to continue to assist with the TRA, and established the CUSA Engineering Consulting Group (CECG). The CUSA was established in 1948, but was later renamed to the Engineering Consulting Company of the Council for International Economic Cooperation and Development (CIECD). In addition, the TRA created the US Aid Technology Team for US aid–related affairs, which was later renamed to 'the International Affairs Team' in 1957 (Wen 2010, 87).

These American advisors would be inspecting TRA development, writing reports, and suggesting proposals. However, they occasionally came into conflict with TRA managers. For example, the previous chief and president of TRA transport services Shu-Hsi Chen mentioned his memoirs, "In the US aid period (1953–1958), Richard C. Morse, the Irish-American advisor, was no longer interested in passenger transportation because of the recession that had hit the American Railways. In addition, he was eager to show his abilities by improving the country's freight transportation, and wanted to push for raising transportation capacity. However, 81% of the TRA's income is profits from passenger transportation, and 19% of its income is profits from freight transportation. Therefore, the efficiency of his options is limited. Finally, the TRA will need to cease taking US aid and develop independently (Chen 1995b. 5, 52)."

Morse's persistence with regard to TRA freight transportation impressed Shu-Hsi Chen. Morse believed that the profits from freight transportation would exceed those of passenger transportation, and was serious about pushing the development of freight transportation. Moreover, based on looking at existing equipment he believed that the TRA was able to achieve the goal of moving 30,000 tons of freight daily. While furthering his interest in developing freight capacity, Morse ignored how crowded passenger transportation had become due to a lack of passenger cars, and impeded TRA passenger car procurement plans (Chen 1995a. 2, 68–70). Shu-Hsi Chen pointed out that Morse ignored the actual needs of Taiwan in favor of introducing Taiwan to 'the American experience', which attached great importance to freight transportation. In other words, the US often made unilateral decisions about these development plans, and seldom queried aid-receiving countries about their options first. This created potential for great differences between such decisions and the actual requirements of aid-receiving countries.

5.2.2 American and Japanese Railway Fact-Finding Delegation in 1953

During the Japanese colonial period, the RD adopted Japanese-style or transplanted Japan Railways technology, especially hardware and cars. The development experience of Japan Railways also became the essential reference for the TRA. Later, as a result of the US aid period, the American experience developing railways would become a similar reference for the TRA—the US, which came to be the ally of the R.O.C. against the P.R.C. and Russia, also owned advanced Railways. American Railway development also influenced the development of the Taiwanese economy, including the TRA itself.

In 1953, approved by the Mutual Security Agency's Mission to China and supported by the Technical Assistance Program, the TRA composed a fact-finding delegation to visit the US and Japan. There were 5 members in the delegation, including the chief engineer Pin-Jhuang Duan, the senior engineer Te-Nien Chen, the transport service chief Shu-Hsi Chen, the engineering chief Hai-Chu Cheng, and the deputy maintenance chief Tsung-Hao Peng. They departed from Taiwan to the US on February 2.

After arriving in the US, the Mutual Security Agency arranged for them to participate in the seventh Railway Transportation Course at the Transportation Institute of American University in Washington D.C. for one month. After finishing the course, the Mutual Security Agency arranged for them to visit the Reading Railroad, the Erie Railroad, and the Chicago, Burlington and Quincy Railroad.

On July 1, 1953, the delegation arrived in Japan. There, Shu-Fang Wang, the Tokyo representative of the Directorate General of Telecommunication (DGT) of Taiwan's Ministry of Transportation and Communications (MOTC), arranged for the Taiwanese delegation to America to visit President of Japan National Railway, Nagasaki Sounosuke. After that, they visited the truck lines and branch lines of Japan's national railway. The delegation returned to Taiwan on August 6, 1953. Upon their return they made a report to the TRA's director general Mo Heng about the railway passenger and freight transports in the US and Japan, and went on to publish the book *Report on Visiting American and Japanese Railways* on October 1953 to construct the essential reference for further TRA development.

The delegation also proposed future developmental directions for the TRA in the aspects of transport services, engineering and maintenance (Taiwan Railway Fact-Finding Delegation 1953, 1–4, 177–182). According to the proposals, the key points and suggestions about power innovation are described as follows:

Prior to the US aid period, a major change the American Railway put through was to adopt diesel-powered locomotives instead of coal-fired ones. In the peak period, there were 64,000 coal-fired locomotives in the US. In 1953, two-thirds of coal-fired locomotives had been discarded and there were only 20,000 left. By this time, locomotive manufacturers had switched to producing diesel-powered locomotives. For perspective on how quickly diesel-powered locomotives were taking over the American railways, there were 510 diesel-powered locomotives in operation in 1939, and by 1953 there were 19,000. Discarded coal-fired locomotives were sold to steel plants as scrap iron at a sell price of $60 USD per ton. The weight of a locomotive was 150 tons on average, and so the scrap price of each locomotive was about $9,000 USD. However, a new locomotive from the US cost more than $70,000 USD, and one from Japan cost more than $50,000 USD. Therefore, the delegation proposed that the TRA could purchase several hundred of these locomotives for their plans to return Mainland China to Kuomintang (KMT) rule in the near future (Taiwan Railway Fact-Finding Delegation 1953, 42–43).

The Japan National Railway did not adopt diesel-powered locomotives due to the lack of crude oil—it would make the Japanese Railway unusable if a war broke out and the oil source was blocked. Also, the Japan National Railway did not adopt steam locomotives because Japan's coal output was not high enough and imported coal was too expensive. Hence, the Japan National Railway focused on railway electrification. By 1953, 10% of 19,000 km national railway had been electrified, and 6000 km of the 7800 km private railway had been electrified. Taking the railway electric showed itself to be advantageous—it had a higher initial cost, but significantly lower costs afterwards. For example, the cost to travel by train using coal was 99 yen per kilometer, whereas the cost to travel by train using electricity was a mere 33 yen per kilometer. Although Japan did not adopt diesel-powered locomotives, Japan National Railway purchased 3 diesel-powered locomotives produced by the Mitsubishi plant for experiments in the Komehara area. The efficiency of these locomotives was lower than those produced in the US, and their prices were higher. The intended purpose of these experiments was the development of technology for export (Taiwan Railway Fact-Finding Delegation 1953, 43–44).

After taking in these findings, the delegation proposed that Taiwan was not ideal for adopting diesel-powered locomotives due to a lack of sources for oil. However, the TRA had an opportunity to procure several shunting or passenger diesel trains for education and investigation. The delegation suggested that, in the process of fulfilling the KMT's plans of retaking Mainland China, the diesel-powered locomotives would be appropriate to the northwest plateau area, which has an output of crude oil (Taiwan Railway Fact-Finding Delegation 1953, 44).

5.2.3 US Aid and TRA Development

The *US Aid Application Review for Taiwanese Railway Construction*, published by the Council for International Economic Cooporation and Development of the Executive Yuan, can be used to show the attitude of the Government and their appraisal of the situation. The book may not fully indicate the timeline of US aid, but it clearly demonstrates the aid's value to the TRA, as follows:

a. Railway Restoration Period (1950–1953): The first priorities were to get the railways operational again, continue transportation, and to ensure safe operation. Since the status of the railway lines impacted traffic directly, the restoration aimed to repair roadbeds, to unblock culverts, to mend wing walls, to replace and to stabilize rotted railroad ties, worn steel rails, and corroded bridges, to fix the machines connecting traffic signals and stations, and so on. Second, the amount of available locomotives, passenger cars, freight cars, and oil cars was not enough to fulfill increasing business demands. The TRA repaired damaged cars and assembled new ones to support urgent requirements. In addition, the station houses, warehouses, platforms, footbridges, sheds, and other facilities were re-evaluated not only for business management but also for the safety of people and goods. In 1949 the TRA made a 4-year plan (1950–1953) to use aid from the US to procure 8 CT270 steam locomotives, 15 diesel engines for passenger cars, 151 covered cars, and 10 Ping-Deng passenger cars, as well as steel bridge materials, paint, railroad ties, steel rails, communication devices, car repair materials, and so on. The restoration was completed quickly because of the efficient use of US aid (Council for International Economic Corporation and Development, Executive Yuan 1964, 2).

b. Power and Car Supplement and Rail and Bridge Maintenance Period (1953–1956): Increasing the transportation capacity is the essence of railway business. In addition to increasing available passenger and freight cars, the TRA aimed to resolve pre-existing issues such as aging train cars, inconsistent rail weight, the loading of bridges being beneath standards, and so on. The TRA also made use of machine tools to better maintain railways and gradually improve loading and unloading freight operation. Since the TRA used the 1954–1956 US aid to purchase diesel engines, diesel passenger and freight cars, railroad ties and base plates, stone crushers, tie tampers, handling equipment, and so on, these improvements were completed on schedule. With these improvements, the capacity of locomotives was increased, and the average speed of trains was enhanced to exceed the highest records from the Japanese colonial period (Council for International Economic Corporation and Development, Executive Yuan 1964, 2–3).

c. Railway Equipment Expansion Period (1957–1960): The European and American railways renovated the way their own trains were powered based on the power sources of their countries. For example, the railways would be switched to electric or diesel power sources. In addition to their original steam locomotives, the TRA purchased 52 diesel-powered locomotives in this period to dieselize their passenger cars completely. Since the signal facilities can impact train safety directly and improve railway traffic systems, traffic capacity and transportation efficiency could be fully utilized. The TRA adopted automatic signal facilities, such as the newest relay interlocking devices and automatic traffic-blocking devices, as well as centralized traffic control systems. Hence, the traffic capacity of each line was enlarged further, and the train delays were also improved. In addition, the TRA procured a lot of steel materials to strengthen bridges. Moreover, the TRA procured steel passenger cars and freight cars, or made conversions to replace the original wooden cars, which were unsafe and had higher maintenance fees. The TRA also expanded railway workshop facilities to manufacture passenger cars and freight cars, as well as to maintain new locomotives. These development plans were implemented successfully through the 1957–1960 US aid period and Development Loan Fund (Council for International Economic Corporation and Development, Executive Yuan 1964, 3).

As shown by information presented in this section, in the interest of the efficient use of US aid, American advisors and the US consulting companies played important roles in examination and implementation that came to impact Taiwanese economic and TRA development. American advisors would inspect the TRA's ongoing development, write reports, and raise proposals. However, the US often made unilateral decisions for these development plans, and there sometimes would be discrepancies between the decisions made and the actual needs of aid-receiving countries.

1953's TRA fact-finding delegation sent to investigate American and Japanese railways was also supported by US aid. This delegation's learning experiences are one way that that American and Japanese experience developing railways became an indispensable reference for TRA construction going forward. The proposals of the delegation contained many suggestions for railway development with regard to and after Mainland China being recovered by the KMT. Generally, the Executive Yuan's Council for International Economic Cooperation and Development tended to acclaim the influence of US aid on the TRA since the US contributed a great deal of material and technical support. However, are the ways in which the US benefitted Taiwan the primary dimension of this situation? Did the TRA provide anything to benefit the US? The following section will describe the issue based on the viewpoint of power innovations.

5.3 The Partial Dieselization Policy and Its Forerunner Policies: Diesel-Powered Cars

5.3.1 From the Argument Regarding Dieselization and Electrification to Partial Dieselization Policy

In 1953, the TRA fact-finding delegation found that Japan had adopted widespread electrification of their railways, and that American railways were making use of diesel-powered locomotives instead of coal-fired ones. Comparison of the railway energy source policies of Japan and the US brought to light important differences from which much could be learned. The delegation proposed that adoption of diesel-powered locomotives was not appropriate to Taiwan, since oil was not conveniently accessible. However, the TRA found that it could procure several shunting or passenger trains for further education and investigation (Taiwan Railway Fact-Finding Delegation 1953, 42–44).

In 1958, TRA's director general, Heng Mo, presented a doubt as to whether dieselization or electrification should be first at the Taiwan Provincial Transportation Development Conference. The conference suggested that the issue should be investigated by the TRA first (Department of Transportation, Taiwan Provincial Government, 1958, File ID: 0640000000006A). The conclusion was that the issue could not be properly discussed at that time, as the TRA had still not investigated it completely. In addition, the TRA mentioned the possibilities of dieselization in the book *Taiwan Railways in the Past 10 Years* (1949–1958) that, "Each advanced country focuses on energy source innovations for their trains, either electrification or dieselization, based on the power sources available. The purposes of this are to pursue larger tractive force and providing a rapid, comfortable, and clean passenger transportation service. Currently, the TRA is not ready to move in the direction of electric train systems, and has instead prioritized diesel. For example, the Fei-Kwai-Che, which provides rail transportation between Taipei and Kaohsiung, makes use of several linked diesel rail cars. Its maximum speed is 95 km/hr, and the duration of going one way from Taipei to Kaohsiung is only 5.5 h. This convenient transportation service is a favorite for many passengers. In addition, diesel-powered locomotives that have been purchased to continue the dieselization of passenger trains are being delivered and put into operation gradually. However, at this time, not all of our steam locomotives should be discarded, since Taiwan outputs coal and our electric power is still insufficient (Taiwan Railway Administration 1959, 3)."

Obviously, power modernization was one of the most important aims of the TRA during the US aid period. Since the two options of dieselization or electrification were the only options, it fell to the TRA to weigh them very carefully. The cost of electrification was too high, and the electrification would not be able to achieve optimal efficiency before all related construction was totally finished. Moreover, the situation across the Taiwan Strait was very tense. If a war broke out, electrified railways could be damaged so as to impact the military (Zhao 1983, 31–33). Conversely, the amount of gasoline in Taiwan was not enough to support railway dieselization as had been done in the US. Therefore, the TRA adopted the policy of switching passenger cars to diesel technology first, since passenger transportation was the major business of TRA.

Though diesel and steam-powered locomotives were both in operation and more diesel cars were being added, it became obvious that it was necessary to transition to electric railways. The TRA adopted a policy to increase the amount of diesel-powered locomotives and to eliminate the steam locomotives as it became feasible to do so. While this was being done as a stopgap measure, the TRA would look to the future and consider how best to achieve the railway electrification.

The minutes for the 6th meeting of the Executive Yuan's Council for US Aid, which took place in 1959, also show the transition policy and the possibility of electrification in the future: "From this year, TRA will adopt diesel-powered locomotives instead of steam. The TRA has ordered 22 diesel-powered locomotives, and will no longer add steam locomotives. In order for the railways to be completely dieselized, the TRA requires about 200 diesel-powered locomotives, which would require 50,000 tons of diesel every year. Although it would be possible for Taiwan to manufacture its own diesel, it would first be necessary to import crude oil. It became necessary for the TRA to refer to the railway experience of Japan and the western European countries in order to investigate how to achieve railway electrification, especially due to electrification of the railways having high initial costs in contrast to significantly lower costs of operation. Were the TRA to start developing electric railways, the purchased diesel-powered locomotives would not be enough to fulfill their needs in the meantime. Therefore, it is concluded that the TRA should also put at least some resources into acquiring more diesel-powered locomotives for transportation services for branch lines and shuttle stations (Yeh 2009, 476–477)."

These meeting minutes show that the TRA made the decision to eliminate steam locomotives. However, from the viewpoint of power source development, Taiwan had no conditions in which to develop electrified railways like Japan, nor diesel railways like those of the US. Therefore, TRA adopted the transition policy of electrification in the trunk lines and dieselization in the branch lines. This way, the TRA would purchase diesel-powered locomotives and distribute them to each line while eliminating steam locomotives as possible. As electric railways could be developed for trunk lines, TRA would then arrange for diesel-powered locomotives in branch lines, the eastern line, and in stations as needed for shuttles.

In 1958, there were more than 200 steam locomotives in the western line. The the average age of a TRA steam locomotive was 33.1 years, and 62.4% of steam locomotives exceeded the standard service life. These locomotives were low safety and high cost. The TRA developed a 15-year phase-out program for old locomotives. This would eliminate 154 steam locomotives. Since they were replaced by new locomotives gradually and were not manufactured in each country, the TRA added 118 diesel-powered locomotives in the second phase of its 4-year CEPD plans, intending to promote electrification in trunk lines within the third phase of these plans (Taiwan Railway Administration 1959, 44).

A flexible and pragmatic policy was used for the TRA's power source innovation. After purchasing its final batch of steam locomotives in 1953, the TRA started to do development based on its new policy. It adopted this strategy, 'partial dieselization', for its transition, and waited for electrification to occur in trunk lines.

5.3.2 Dieselization Harbinger–Diesel-Powered Cars

There are two basic kinds of dieselized locomotives: the internal combustion passenger car and the diesel-powered locomotive. An internal combustion passenger car, called a rail car, is a passenger car which can be equipped with an internal combustion engine. Each internal combustion passenger car can be driven independently when equipped with an engine. Also, several linked internal combustion passenger cars can be used to pull some trailers lacking engines.

This method of operation is special, as usually a train is composed of several non-powered cars pulled by a powered locomotive. Because of its smaller size, lower operating costs, higher speed, and other advantages, it was easy to dispatch internal combustion passenger cars as shuttle trains by adjusting their transportation frequency based on actual needs. Hence, they are very appropriate to Taiwan, which is small in area but densely populated, and has great demand for passenger transportation (Taiwan Railway Administration 1959, 22). The history of internal combustion passenger cars in Taiwan can be traced back to the 35 gasoline railcars (GA) produced by the Japanese companies Nippon Sharyo and Kawasaki Rail Cars during colonial Japanese rule in the 1930s. During the time of the Pacific War, most of the gasoline railways were damaged due to air bombing and a lack of oil, and those available suffered from lack of maintenance. After the war, the Railway Administration Council took over 35 western-line gasoline trains classified as small, medium, and large types. Except for 5 scrapped gasoline trains, the 11 small trains kept the same bodies and bogies, and were not equipped with an engine because of the costs. They were then defined as DR1000 and DR1010 and transferred to use as passenger cars. The 19 remaining medium and large train cars were each equipped with either a new gasoline engine or a repaired older engine. However, the engines of these gasoline railcars were too old to maintain. Also, they failed often, and it was difficult to keep them supplied with necessary supplemental equipment.

The TRA applied the 1954 US aid to purchase fifteen 200-horsepower Cummins diesel engines, and then installed them on thirteen large gasoline railcars. In 1956, the TRA went on to apply its budget to the purchase of seven 160–horsepower DMH17 diesel engines (Japan National Railway Standard) , and then installed them on 6 medium gasoline railcars. From this point on, all of the medium and large gasoline railcars were now retrofitted for diesel, and their car numbers were changed from GA2100–2400 to DR2100–2400, in which the DR was the abbreviation of 'Diesel Railcar' (Wen 2010, 95). In 1954, the DR2300–2400 diesel passenger cars were put into use between Taipei and Taichung by Fei-Kwai-Che. At this time on this route, the duration of a one-way trip was 3 h—a half-hour faster than a steam locomotive's one-way trip duration.

At the same time, the TRA trained technicians had built internal combustion maintenance systems in preparation for transitioning to diesel power. Since the performance of these diesel passenger cars was good, the TRA applied for US-aid to purchase eighteen 300-horsepower large diesel cars and 4 trailers from Japan's Tokyu Car Corporation in 1956 to fulfill their needs for passenger transportation. These diesel cars and trailers served between Taipei and Kaohsiung under Fei-Kwai-Che in 1956 and 1957 respectively. The duration of a one-way trip was 5.5 h, and they were a popular method of transportation (Council for International Economic Corporation and Development, Executive Yuan 1964, 22–23).

The development of the eastern line was similar to that of the western line. First, the TRA kept the original steam locomotives and converted engines for additional support during the most difficult stages. The TRA then purchased diesel-powered cars to raise transportation capacity, though the eastern line's progress was slower than that of the western line. In 1956, TRA procured ten diesel engines, including seven more 200-horsepower American Cummins engines and three 125-horsepower Japanese Hino engines. The 7 Cummins engines were installed in three original large gasoline railcars and also in 4 additional diesel-powered cars equipped with new bodies, and the 3 Hino engines were installed in 3 original medium gasoline railcars. After that, there was a total of 14 internal combustion passenger cars, including 10 diesel-powered cars and 4 original small Isuzu gasoline railcars, in use on the eastern line. Since each of American Cummins diesel cars was powerful enough to pull 2 trailers, they were operated for passenger transportation services. Since 1957, the capacity of the 14 internal combustion passenger cars was not enough to provide transport to the increasing numbers of passengers. TRA converted 2 cars that were originally 186-horsepower White engine cars on the western line into 2 medium cars for the eastern line, and installed the 2 converted Hiro engines in 2 small cars. Since these converted cars could pull a trailer, the total transportation capacity was enhanced and the dieselization had made further progress. With dieselization, the speed of passenger transportation was also enhanced from 60 to 70 km/hr. The one-way duration of a trip from Hualien to Taitung was shortened to 3 h and 45 min. This enhanced turnover ratio caused more transportation services to be available and raised total transportation capacity (Taiwan Railway Administration 1959, 75).

In 1959, US aid was used to purchase 4 diesel-powered cars which were equipped with American Cummins engines, with bodies constructed by the Japanese Tokyu Car Corporation. The Central Trust Bureau processed this procurement in 1960, and the 4 engines were delivered to the TRA in March of 1961 before being assembled and put into operation on the eastern line. In addition, the application for 2 diesel-powered cars to be purchased with US aid in the fiscal year of 1960 was finalized that same year.

The budget was not enough for the original plan, by which the TRA would have purchased 2 complete cars. In light of this, the TRA and the joint American advisor, a man surnamed Bezier, modified the plan, instead choosing to purchase materials and construct the cars in Taiwan. The Central Trust Bureau purchased materials that it was necessary to order from overseas, while locally available materials were purchased by the TRA. The Taipei Railway Workshop manufactured the car bodies, and the 2 railcars were then assembled for the eastern line. The manufacture started in September of 1961, and the 2 cars were finished and operational on the eastern line in January and February of 1962 respectively.

All of the 6 US aid diesel-powered cars mentioned herein were designed by the TRA. They were popular with passengers because of advantages such as their bodies being longer than the eastern line's original diesel-powered cars, their being able to provide more seats, modernized internal design, softer lighting, better overall performance, and no swaying during operation. These 6 cars were responsible for express services on the eastern line. The distance between Hualien and Taitung is 170.4 km, and the duration for this kind of car to make the trip was a mere 3.5 h–much less than the 5 h and 30 min duration of the original locomotives in use prior to 1956. They made a great contribution to the improved operation of the eastern line (Council for International Economic Corporation and Development, Executive Yuan 1964, 61–62).

5.4 The Innovation of Locomotives

5.4.1 From Maintenance to Expansion

The *Report on Transportation and Communication in the Past 3 Years* mentions, "The age of more than half of existing TRA locomotives in 1952 was over 30 years, and for some was over 50 years. Moreover, 54 locomotives had been damaged in the early post-war period because of air bombing during wartime or were in disrepair for years. After 28 damaged locomotives had been repaired from

1950 to 1952, together with repaired locomotives from years ago they had had all been repaired and were operating normally. Also, the TRA purchased 5 large locomotives to significantly raise its transportation capacity (China Communications Construction Association 1953, 755)."

To meet transportation demand in the early post-war period, the TRA repaired and maintained its locomotives actively, completing this restoration project in 1952. In total, the TRA repaired 95 damaged and broken locomotives, produced 6 new locomotives by scrapping and reintegrating parts from 31 locomotives, and purchased 10 freight and 8 passenger locomotives. Although the amount of locomotives in use in 1958 was similar to the amount in use in the early post-war period, the transportation capacity had been doubled (Taiwan Railway Administration 1959, 19).

Until the US aid period began in 1950, the TRA applied US aid to purchase 5 DT650 steam locomotives (DT683-DT687) from Nippon Sharyo and the Japanese company Mihara in 1951 for increasing transportation demands. In 1953, the first year of the first 4-year CEPD plan, the TRA applied for US aid to purchase 8 CT270 steam passenger locomotives (CT277-CT284) from the Japanese Hitachi and Kawasaki companies. These 8 locomotives were delivered to Taiwan in two sets and assembled in the Taipei Railway Workshop. Before introducing diesel-powered locomotives, the CT270 locomotives were the major cars for railway transportation. They were intended to raise the transportation capacity and speed for long-distance passenger transportation. The standard daily transportation distance of passenger locomotives was raised from 300 to 360 km (Taiwan Railway Administration 1987, 66; Chou, 1995, 130; Council for International Economic Corporation and Development, Executive Yuan 1964, 22). However, these steam locomotives were still inferior to diesel-powered locomotives (Chang 1991, 60–61).[3] After purchasing CT270 locomotives in 1953, the TRA did not purchase any steam locomotives and directed its efforts towards dieselization. Since these purchased steam locomotives in the early post-war period were mostly Japanese-style, it showed the situation that the TRA's continued development still depended on Japan for support. This could perhaps be called a cultural holdover from Taiwan's colonial Japanese era.

[3]Referred to Chang, Chun (1991), *Theory and Practice of Railway Transportation* (pp. 60–61). Taipei: Taiwan Business, "Although steam locomotives are easy to maintain and repair (and technically easier to handle), the following are the major disadvantages comparing to diesel-powered locomotives: 1. Low tractive force and speed since a large stream locomotive can trail 2,000–3,000 tons and a diesel-powered locomotive can trail 7500–10,000 tons; 2. Very low thermal efficiency since a steam locomotive operates at 6% and a diesel-powered locomotive operates at 22%; 3. Additional drivers are required in linked steam locomotives for trail and only 1 driver is required in a diesel-powered locomotive linking with others; 4. High maintenance cost; 5. Large impact force on rails; and 6. Air pollution making customers uncomfortable.

5.4.2 Diesel-Powered Locomotive Procurement

In the face of increasing transportation demands, steam locomotives were not only restricted by their traction and speed, but also experienced more frequent delays and created more soot, making it increasingly difficult to fulfill passengers' transportation demands. It was necessary to improve the equipment to further develop the business, as well as to attract passengers and clients for the shipment of goods. A diesel-powered locomotive is a locomotive equipped with a diesel engine that in turn drives a generator. From there, the current will be converted to D.C. through the filter and rectifier, and the motor on the wheel axle will drive the locomotive forward. Based on this operation, this kind of diesel-powered locomotive is very powerful but its speed is easier to control (Su 2009, 57). The introduction of diesel-powered locomotives showed that the TRA was entering a new era of transportation service. However, the procurement of this new equipment was full of twists and turns.

With the second phase of CEPD plans taking place from 1957 to 1960, the TRA applied for US aid to purchase 52 diesel-powered locomotives for complete passenger car dieselization. These purchased diesel-powered locomotives contained 40 large units (10 Japan Hitachi RO and 30 US GM R20) and 12 small units (American GM S200). They were delivered to Taiwan and put into operation in 1960 and 1961, respectively. These diesel-powered locomotives would be used for transportation services of the express, major passenger trains on the Yilan line and Tamsui line to raise speed and shorten overall travel time. Also, the TRA put a new, more efficient train timetable into use. Improved operations established good transportation and inspection systems, and were an important step in total transition to diesel power (Council for International Economic Corporation and Development, Executive Yuan 1964, 23).

The first batch of 10 diesel-electric locomotives was purchased using US aid for the fiscal year 1957. This was done using the American diesel-electric locomotive as a model for what they wanted to obtain and how they wanted it to work, and using the process of an international, public tender, upon which different companies from different countries could bid. The TRA originally expected the tender to be obtained by the technically sophisticated US manufacturers, but the Japanese company Hitachi won the tender by making bids so low that one Japanese car was $300,000 USD cheaper than its GM equivalent. However, the TRA was doubtful of Hitachi's lack of experience in manufacturing diesel-electric locomotives, and the Mutual Security Mission to Taiwan also shared the TRA's opinion and agreed to the TRA's suggestion that Hitachi be disqualified from the bidding war due to non-compliance with the tender's specifications. After the procurement case was sent back to the US Mutual Security Agency for review, the US Government, after political and diplomatic considerations, proposed that the TRA go with the results of the bidding war and make the purchase from Hitachi. Hitachi, understanding the TRA's doubts, offered 17 specific technical assurances, as well as 2 additional diesel-powered locomotives. After the start of production, Hitachi ran into problems, even needing to re-manufacture to address non-compliance with specifications. Consequently, the order delivery schedule was delayed, and the TRA's timetable for transitioning to diesel was adversely affected. Greatly dissatisfied with Hitachi, the TRA indicated that it would not purchase diesel-electric locomotives from Hitachi again in the future (Chiu 2010, 20).

Before the TRA purchased diesel-powered locomotives under the first phase of Development Loan Fund projects, it made a survey report comparing GM's large-scale locomotive with the Hitachi's locomotive in areas such as design, manufacturing, materials, difficulties in inspection and maintenance, the amount of parts used, fuel consumption, availability, frequency of malfunction, and so on. The final report, called *Survey On the Use of Diesel-Electric Locomotives Manufactured by Hitachi and GM*, gave an extremely low rating to Hitachi's locomotives. Likely with this as a pretext, the TRA didn't buy locomotives from Japan. When the TRA borrowed US $3,200,000 from the Development Loan Fund to purchase 12 diesel-powered locomotives and Centralized Traffic Control signal devices for the sea line from Hsinchu to Taichung, it was decided to use a non-public tender, because the TRA intended to make the purchase from firms with extensive experience in manufacturing. It compared price quotations from GM, GE and ALCO and opened the tender in the US, where GM won the contract (Taiwan Railway Administration, 1970, File ID: 36-05-043-003; Chiu 2010, 21). Finally, the TRA ordered 11 diesel-powered locomotives from GM, and contracted that they should be delivered in two sets. The first batch contained 4 large diesel-powered locomotives which were GM G12, the standard type for export (Anonymous 1960a, 2, 1). They were delivered to Taiwan on April 3, 1960 (Anonymous 1960b, 4, 6). The second batch contained 7 GL8 small diesel-powered locomotives. They were delivered to Taiwan in installments starting June 18, 1960 (Shou 1960a, 7.1, 2).

Although the TRA had signed the contract with Hitachi for diesel-powered locomotives before they had signed one with the American GM, these Hitachi diesel-powered locomotives were delivered to Taiwan much later. 5 of the 12 diesel-powered locomotives arrived at Keelung Harbor on July 10, 1960. Each diesel-powered locomotive weighed 84.5 tons. Their maximum traction was 20.25 tons, and maximum speed was 100 km/hr. Each was equipped with a MAN engine. The MAN engine, the so-called Hitachi-Man, was designed by Germany and manufactured in Japan since Hitachi purchased the patent. It was a V-type, 4-stroke engine, and had 12 cylinders. Its continuous output was 1,420 horsepower and its maximum output was 1,560 horsepower (Shou 1960b, 7.16, 2). Although the horsepower of the Hitachi diesel-powered locomotives was higher than that of the American GM locomotives, the Hitachi engine, manufactured based on Japanese and German technology, malfunctioned more often and was harder to maintain. Malfunctioning and difficulty of maintenance caused locomotives to be operative less than 50% of the time (Taiwan Railway Administration 1987, 70).[4] The failure of dieselization in the beginning was in large part due to poor judgment on the part the U.S.

In February 1960, the TRA proposed a transportation enhancement program, for which it obtained an additional $5.9 million USD from the Development Loan Fund to purchase 31 diesel-powered locomotives (Yeh 2009, 902). Based on the TRA's firm request a clause was added to the contract restricting the tender to being contracted only by US firms. The TRA had to negotiate the contract for a long time to push for this clause's inclusion. After the contract was signed, the Taiwanese Government instructed the TRA to compare quotations from 3 American firms and had the New York Office of the Central Trust Bureau to compare the quotations provided (Chiu 2010, 26). Though they were at first intended just to make the comparison, they were also ordered to handle the negotiations and purchase from GM.

Since all European, Japanese, or even other American firms had no chance to bid for the tender, a dispute arose over the purchase, this time over the selection of GM with no other companies having been given a chance. The TRA argued that it had adopted the opinion of Mr. Moorman, the Chief Engineer of the Development Loan Fund. He thought it very uneconomical to use several brands of diesel-electric locomotives. This was intended to avoid disputes like that which had happened with the US interference in the last purchase, which arose from objections raised by the manufacturers to the TRA's locomotive specifications–which was complicated by the fact that these specifications had already approved by the Development Loan Fund. In brief, the TRA argued its points on the grounds of "power standardization" and "avoiding objections". The move caused instead greater doubt, and it was spoken of whether or not the commission was being given to GM due to kickbacks to KMT officials (Taiwan Railway Administration, 1961, File ID: 36-05-043-006).

In 1960, the TRA originally intended to apply to the Development Loan Fund for the third time in order to to purchase 34 diesel-electric locomotives–however, they found that the soaring price of diesel at the time would cause diesel-electric locomotives to be a source of huge losses. Because of this, the TRA thought it would be unable to repay the loan principal and interest until the oil price issue was reasonably solved. After that, the TRA and other Taiwanese organizations tied to the oil market spent a long time focused on solving the problem of oil prices.

In early January 1963, the TRA submitted a project application to the US Agency for International Development's Aid Mission to China. The Mission, in return, indicated that it would need to evaluate the TRA as a business. For example, "how the benefits of investment with continued loans will be effected by the current financial situation", "to what extent is dieselization to be developed", "how the

[4]Referred to the Taiwan Railway Administration, *Taiwan Railway Centennial* (pp. 70), "In 1971, the Taiwan Railway Administration (TRA) demolished the R1 engine and installed the American G.M. engine 645E. It was found to be useful after testing. After that, the TRA reconstructed the engines of the other 6 locomotives, completing 7 in total. The other 6 locomotives were scrapped since they were useless and it was hard to get their components.".

benefits of diesel-powered locomotives compare with those of steam locomotives", and similar queries were put forth by the Mission. The TRA revised and presented its project to the Executive Yuan's Council for US Aid to be forwarded to the US Aid Mission.

In response, the TRA were told by the US Agency for International Development's Aid Mission to China, "The World Bank's experts are currently studying the needs of the overall operation of the railway as well as other means of transport, so this application for a loan will not be considered for the time being (Taiwan Railway Administration, 1963, File ID: 36-05-043-009)." The US side rejected the case and the dieselization policy was stalled for some time. At the same time, the TRA's foreign funding was starting to be changed from US aid to being provided by the World Bank. This signaled the TRA's foreign funds and technology would enter a new stage (Table 5.2).

5.4.3 Power Source Innovation Synergy

Power source innovation during the US aid period had many significant effects in the TRA's operation. First, the TRA promoted the dieselization of the eastern line by converting engines starting in 1956, and finished the dieselization process in May of 1961. Dieselization not only improved the comfort of travel but also reduced the cost (see Table 5.3) (Council for International Economic Corporation and Development, Executive Yuan 1964, 61–62). Second, the average age of active locomotives was reduced to ensure better transport capacity as well as safety. Since the 52 purchased diesel-powered locomotives were put into operation and old locomotives were being eliminated, the average age of locomotives in operation was reduced from 33.1 years in 1958 to 27.2 years by 1963. Third, the transport capacity was increased and the locomotive mileage was increased naturally because the newly purchased railcars had excellent performance and were suitable for high-speed operation. Since new locomotives performed better and raised overall efficiency, the average daily locomotive mileage was increased from 38,966 km in 1954 to 53,579 km by 1963. Going forward, the travel time was shortened and the car rotation rate was increased because the newly purchased railcars had higher horsepower and were suitable for high speeds. The greater horsepower and suitability for high speed of new locomotives shortened travel time and enhanced the turnover rate of locomotives, passenger cars, and freight cars.[5] Finally, the mileage

[5]1. The one-way time cost of a diesel-powered Fei-Kwai-Che train ride between Taipei and Kaohsiung was 5 h and 25 min; 2. The one-way time cost of limited express train ride between Taipei and Kaohsiung was shortened from 7 to 6 h; 3. The one-way time cost of the Ping-Deng express train between Taipei and Kaohsiung was shortened from 8 h to 6 h and 40 min; 4. The one-way time cost of a train ride by ordinary railcar between Taipei and Kaohsiung was shortened by 1 h on average; and 5. The one-way time cost of a freight train trip was shortened by 2 h on average.

Table 5.2 TRA purchased/produced locomotive and diesel-powered car list (1945–1965)

Car	Item					
	Type	Horsepower	Weight (by ton)/Engine type	Max speed (km/hr)	Manufacturer	Year of manufacture
Diesel-powered Car	DR2000	56	American Cummins NHHB-600	70	Japan Tokyu	1947
Diesel-powered Car	DR2050	56	American Cummins NHHB-600	70	Japan Tokyu	1947
Steam locomotives	DT650				Nippon Sharyo, Mihara Cars	1951
Steam locomotives	CT270		11.5	100	Japan Hitachi, Kawasaki Cars	1953
Diesel-powered Car	35DR2500	300	American Cummins	110	Japan Tokyu	1955
Diesel-powered Car	35DR2600	300	American Cummins	110	Japan Tokyu	1957
Diesel trailer	35DR2650			110	Japan Tokyu	1957
Diesel-powered Car	LDR2200				Taipei Railway Workshop, Hualien Railway Workshop	1956
Diesel-powered Car	LDR2300				Japan Tokyu, Hualien Railway Workshop	1961–1965
Diesel-electric Locomotives	R20	1425	78	100	American GM	1960
Diesel-electric Locomotives	S200	950	65	100	American GM	1960
Diesel-electric Locomotives	R0	1420	84.5/German Man	100	Japan Hitachi	1960

Source
1. Hsiao et al. (1994), pp. 1–12
2. Council for International Economic Corporation and Development, Executive Yuan (1964), pp.22–26

Table 5.3 Comparison of eastern railway passenger train income and cost (unit: NTD)

Item	Year							
	1955	1956	1957	1958	1959	1969	1961	1962	1963
Income per passenger per kilometer	0.1430	0.1490	0.1690	0.1783	0.1819	0.2277	0.2422	0.2558	0.2542
Cost per passenger per kilometer	0.2052	0.2086	0.2344	0.2048	0.1867	0.2239	0.2520	0.2837	0.2914
Gain or loss	−0.0622	−0.0596	−0.0644	−0.0265	−0.0048	0.0038	−0.0098	−0.0279	−0.0372

Source Council for International Economic Corporation and Development, Executive Yuan (1964), p. 61

per car per day was heightened markedly because of improved power and increased business volume. Since the diesel-powered passenger cars and locomotives were adopted, issues with soot were also reduced and travel time was considerably shortened. The transportation services were more convenient and both passenger and freight transportation were improved (Council for International Economic Corporation and Development, Executive Yuan 1964, 28–30).

According to the discussion in this section, in the early post-war period, the TRA not only aimed to repair damaged locomotives, but also to purchase steam locomotives based specifically on their actual requirements. Since these purchased steam locomotives in the early post-war period were mostly Japanese-style, it showed that the TRA's development still depended on Japan, which could be viewed as a cultural remnant of the Japanese colonization of Taiwan. After purchasing CT270 in 1953, the TRA did not purchase any further steam locomotives and aimed itself fully towards dieselization. The TRA's procurement of diesel-powered locomotives gave Japan an opportunity to attempt to compete with the US. Although Japanese firms won the first tender because of their low bid, as well as goodwill and support from the US, the high malfunction rate of Japanese locomotives caused the TRA much distress and loss, and the Japanese firms could not win any other tenders. Except for the first 12 diesel-powered locomotives, the TRA purchased diesel-powered locomotives from the American company GM because of its extensive experience in manufacturing. In fact, based on the influence of US aid, the TRA American advisors' recommendations, and the machinations of politics and diplomacy, the original TRA with its typically Japanese style went entirely towards a series of diesel-electric locomotives made in the US. Although the TRA's decision seemed to favor GM excessively (for the reason of power standardization) and caused a lot of conjectures and disputes, the partial dieselization policy led to significant effects from the perspective of innovation in power source.

5.5 Conclusion

The railway research can be classified under the fields of technical research as well as business research. Both of these were deemed important by the US and Japan. The US centralized its railway research to investigate related technologies. The Association of American Railroads focused on railway technology research, and within that set up engineering, maintenance, transportation, and other departments. Japan National Railways set up its own Railway Technology Institute. Although its building and equipment are not better than those of the US, their research projects and contents are not inferior. The TRA did not set up any research institution like those run by the Japan Railway and Manchu Railway before the war, and instead set its business research and development team to investigate railway business and didn't do technical research (Taiwan Railway Fact-Finding Delegation 1953, 26). Hence, it is hard to make breakthroughs and TRA still depends the oversea companies. In the *Economic Analysis of Post-war Taiwan*, Liu, Jing-Qing said, "In the development of the structure of Taiwan's dependence on US aid and Japanese trade, the Taiwanese economy was fixed in this system of dependence on the US and Japan. Taiwan's economic necessities, as well as military and political necessities (particularly under the Cold War system), influenced Taiwanese politics decisively. At the same time, the new colonial dominance of the US and Japan over Taiwan became real (Liu 1994, 351)."

On the surface, the economic links between Taiwan, America, and Japan may be as the same as Liu's interpretations. However, the detailed operations of the system and structure do not, in truth, reach the extent of "domination". Moreover, without full American support in politics, diplomacy, military, and economics due to the Korean War, Taiwan may not have developed as well towards its future economic prosperity. In the *Dependent Development of Taiwan*, Chen, Yu-Hsi proposed a more fair description, "When the US Agency for International Development (USAID) began intervening in and influencing Taiwan's economic development and progress in early 1950's, what it considered was obviously long-term global economic relationship strategy. … The operation by which US capital entered Taiwan aimed to establish a strong coexistent relationship linking Taiwan's economy, the global system, and the American economy (Chen 1995c, 69–70)."

In the early post-war period, a fast restoration was made by the concerted efforts of the Nationalist Government's elite cadre leaders at the top, retained Japanese management talent in the middle, and the skillful Taiwanese employees at the bottom. However, the whole railway business was still out of order. The physical normalization and progress of the TRA came to be developed in the US aid period. Generally, the Taiwanese government trended towards highly praising the influence of the US aid on the TRA. However, when we investigated the process of power source innovation, we found that the TRA also paid higher costs due to the poor judgment of the US, and the US benefitted in various ways including loan interest. Based on a perspective favoring diesel-powered trains, the TRA was not decolonized and favored Japanese firms excessively—however, most engines were American Cummins engines. Based on the viewpoint of changes to locomotives, they were developed from Japanese colonization to American colonization, and GM enjoyed market dominance. It showed that the TRA's power innovation in the post-war period focused on Japanese diesel-powered cars and American locomotives. Japanese manufacturers had not been fully expelled due to their less advanced technology in large part due to America's political and diplomatic considerations, and a delicate division of labor was formed. It seems that there was little to no progress in technical research, development or self-manufacture. Under the US international political layout in East Asia after the Korean War, the TRA only managed to move its dependence from solely Japan to both America and Japan.

It's worthy of consideration that the relationship between Taiwan, the US, and Japan in this period is formed of cooperation, dependence, and the issues formed by Taiwan obtaining foreign capital and suffering invasion of its capital. If a country wants to develop itself but lacks sufficient funds and technology, it is necessary to cooperate with or depend on other countries. Since the TRA did not attach importance to technical research and development, and expected the "Mainland China Recovery" to take place, it naturally cooperated with the US and Japan (which were the anti-communist alliance) to complete its technological innovations. Taiwan's development conditions were different from the United States' as well as Japan's. The partial dieselization policy adopted by the TRA had its own unique features and necessities, and met Taiwan's phased demands adequately at the time. In this process, the US well exceeded Japan in influence over Taiwan. Also, the US aid operations were highly beneficial for the US firms involved. The cooperation structure of Taiwan, the US, and Japan can be observed clearly in the TRA's development in the US-aid period. This structure's influence can clearly be seen in the long-term development of the TRA.

Acknowledgments This work was supported by Ministry of Science and Technology, Taiwan, R.O.C. Grant Number 105-2410-H-305 -056 -.

Primary Sources

4-year CEPD plan (1957–1959) (Bound Volume) (1958). In *Archives of Department of Transportation, Taiwan Provincial Government*, File ID: 064000000006A. Preserved at Academia Sinica.

American advisors' inspection report (1952). In *Archives of Taiwan Railway Administration, Department of Transportation, Taiwan Provincial Government*, File ID: A31518000 M. Preserved at National Archives Administration.

Council for US aid official document in February, 1957 (Tai Mei Ji (46) Word 0657 Letter) Meeting Minutes of Cross-Island Highway Construction Conference of Council for US Aid and Veterans Affairs Commission (1956), File ID: 064000000964A. Preserved at Academia Sinica.

Taiwan Railway Administration Equipment Expansion Plan, Sea-line centralized traffic control signal devices operators (1964–1970). In *Economic Department Archives*, File ID: 36-05-043-003. Preserved at Institute of Modern History, Academia Sinica

Taiwan Railway Administration dieselization plan—the procurement of 31 diesel-powered locomotives (1958–1961). In *Economic Department Archives*, File ID: 36-05-043-006. Preserved at Institute of Modern History, Academia Sinica.

Taiwan Railway Administration, The 34 diesel-powered locomotives (1950–1963). In *Economic Department Archives*, File ID: 36-05-043-009. Preserved at Institute of Modern History, Academia Sinica.

References

Anonymous. 1960a. First batch of 4 diesel-powered locomotives will come in next month, *Business Communication*, No. 82(1).

Anonymous. 1960b. 4 diesel-powered locomotives arrived at Keelung harbor by an American ship yesterday, *Central Daily News*, 6. (April 4, 1960).

Chang, Chun. 1991. *Theory and practice of railway transportation*. Taipei: Taiwan Business.

Chen, Shu-Hsi. 1995a. Chen, Shu-Hsi memoirs 2: chief of Taiwan railway transportation service for 13 years. *Chinese and Overseas Magazine* 57–2: 66–75.
Chen, Shu-Hsi. 1995b. Chen, Shu-Hsi memoirs 5: president of Taiwan railway for 5 years. *Chinese and Overseas Magazine* 57–5: 52–59.
Chen, Yu-Hsi. 1995c. *Dependent development of Taiwan.* Taipei: RenJian Publishing Co.
China Communications Construction Association. 1953. *Report on transportation and communication in the past 3 years.* Taipei: China Communications Construction Association.
Chiu, Hung-Chih. 2010. The 50th anniversary of Taiwan railway locomotive dieselization. *Biographies* 94–4: 19–29.
Chou, Hsiu-Huan. 1995. *US aid archives in the early postwar period, Volume 1, Military assistance 1.* Taipei: Academia Historica.
Council for International Economic Corporation and Development, Executive Yuan. 1964. *Review on US-aid applications 10–review on US-aid applications in Taiwan railway construction.* Taipei: Council for International Economic Corporation and Development, Executive Yuan.
Hsiao, Hui-Huang, and Yue-Yi Guo. 1994. *Taiwan railway trains 83.* Taipei: Taiwan Railway Administration.
Kerr, George H. 2014. *Formosa betrayed.* Taipei: Taiwan Association of University Professors.
Liu, Jing-Qing. 1994. *Economical analysis of post-war Taiwan.* Taipei: RenJian Publishing Co.
Miller, Merle. 1974. *Plain speaking: an oral biography of Harry S.* Truman: Putnam Publishing Group.
Shou, Jyun-Ren. 1960a. Introduction to the structure and characteristics of the second batch of assembled diesel-powered locomotives unloaded from ships, *Business Communication*, No. 91 (2). (July 1, 1960).
Shou, Jyun-Ren. 1960b. Introduction to the structure and characteristics of the first batch of assembled Hitachi diesel-powered locomotives unloaded from ships, *Business Communication*, No. 92(2). (July 16, 1960).
Su, Chao-Hsu. 2009. *The illustrated handbook of Taiwan railway stock.* Taipei: RénRrén Publishing Co.
Taiwan Railway Administration. 1959. *Taiwan railway in the past ten years.* Taipei: Taiwan Railway Administration.
Taiwan Railway Administration. 1987. *Taiwan railway centennial.* Taipei: Taiwan Railway Administration.
Taiwan railway fact-finding delegation. 1953. *Report on visiting American and Japanese railways.* Taipei: Taiwan Railway Administration.
Takeharu, Sasamoto. 1992. Industrial development process. In *The formation of international machining base: Taiwan's industrialization*, ed. Takao Taniura. Taipei: RenJian Publishing Co.
Tsai, Lung-Pao. 2007. Technology transfer in the colonies - case study of railway staff in Taiwan Governor-General's Office, *Corporate activities and role of financial markets in Asian economic development: history and present.* ed. Hirohiko, Shinbo. Osaka: Osaka Sangyo University Asian Community Research Center.
Wen, Wen-Yu. 2010. *The study of railway history in postwar Taiwan:in the period of Mo-heng (1949–1961).* Taipei: Graduate Institute of Taiwan History, National Chengchi University.
Yeh, Hui-Fen. 2009. *Chen, Cheng's official career archive selection: meeting minutes of council for US aid.* Executive Yuan, Taipei: Academia Historica.
Zhao, Jie-Qian. 1983. *Transportation development strategy.* Taipei: Linking Publishing Co.

Chapter 6
A Comparison of Railway Nationalization Between Two Empires: Germany and Japan

Ayumu Banzawa

Abstract This article compares the failure of nationalization in the German Empire in the 19th century with Japan's earlier railway nationalization; that is, the placement of the main railway lines under national governmental management. Several comparative studies by Japanese economic historians have been made on railway nationalization in Germany and Japan. These studies have focused on the historical stages of capitalism, mainly from a Marxist perspective, and were centered on the concept of the "Bismarckian nationalization" of Japanese railways. It may be helpful to compare the two railway nationalization programs by closely examining the negotiations between local German states and the institutions involved in German railway centralization: the Imperial Railway Bureau and the Prussian Railway Central Bureau. Our inquiry into the official documents (mainly those of the Prussian Kingdom) about railway nationalization in the German Empire shows that the particular obstacles to German railway nationalization had many facets and that the establishment of a central administrative office was technically difficult given the federal system in place. These findings indicate that (1) an accurate subject for a historical comparison with Japanese railway nationalization is not the German national railway integration by the imperial government, but rather the official Prussian purchase of railways, and (2) organizational trial and error in the centralization of the technological railway administration in Germany was identical to Japan's due to the common reaction to the problems with the railway systems at the turn of century.

Keywords Germany · Railway · Nationalization · Administration · Comparative history

A. Banzawa (✉)
Graduate School of Economics, Osaka University, 1-7 Machikaneyama Toyonaka, Osaka 560-0043, Japan
e-mail: banzawa@econ.osaka-u.ac.jp

M. Sawai (ed.), *The Development of Railway Technology in East Asia in Comparative Perspective*, Studies in Economic History,
DOI 10.1007/978-981-10-4904-0_6

6.1 Introduction

Many modern Japanese researchers are interested in the comparison of experiences between Japan and Germany (Europe). The awareness of a comparison between the experiences of the nationalization of the railway industry began to grow on the Japanese-side from early in the 20th century (i.e., at the end of the Meiji Period).

Gotō Shinpei, a statesman of the Empire of Japan from the beginning of the 20th century to the eve of the Great Depression, was the first director of the Japanese Agency of Railways (*Tetsudō-In*) and of the South Manchuria Railway Company (the so-called *Mantetsu*). He wrote the following in a pamphlet of the *Mantetsu* in 1909:

> …public opinion over railway nationalization, establishment of a new Imperial Agency of Railways, and enactment of railway special account has been in a ferment at this moment, from what I have heard. While probably this issue was a very difficult problem to solve even for the Prince Bismarck, … today's status of our Empire is the favor on which we worked over and carried out resolutely the projects on which each country hesitates to determine.

Here, Gotō referenced the Meiji restoration, Constitution Promulgation, Treaty Revision, and the Sino-Japanese War and the Russo-Japanese War as examples and continued:

> And the railway nationalization also has been carried out by us (*Minami Manshū Tetsudō Tōa Keizai Chōsa Kyoku* 1909, 1).

Following these, Gotō pointed out the failure of railway nationalization in the German Empire by Otto von Bismarck (who Gotō referred to as the "Great and lucky Prince B.") because of his opposition of the *Reichstag*, the *Bundesrat,* and the Prussian Parliament in the 1870s. According to Gotō, there would have been no financial difficulties of the German empire "today", (in the beginning in 1900s), if Bismarck's railway nationalization were carried out. Observing the business turnover of the Prussian State Railway, the largest German state railway, Gotō regarded it as a problem that the finances of the Prussian State Railway were not separate from a general account (*Minami Manshū Tetsudō Tōa Keizai Chōsa Kyoku* 1909, 1–9).

It can be said that such an assessment of the German experience by Gotō was to a certain degree correct as far as it reflected historical processes. When his remark was based on an assessment of the conditions of German railway industry in the pamphlet, Gotō's nationalistic tone praised the success of the nationalization in Japan.

But it is also clear that there were some problems with Gotō's comparison. Those will be investigated in what follows.

First, Gotō ignored differences in the meaning of "railway nationalization" between Japan and Germany. There are various realities of railway nationalization that relate to the degrees of state ownership and of national unification. By Gotō, and later some Japanese railway historians who tackled the character of Japanese railway nationalization in the light of the model of so-called "Bismarckian nationalization" in Imperial Germany (Yamada 2001, 185–197), there seems to be some confusion about the concepts. In Japan, the railway nationalization was only the purchase of private railway enterprises and the state ownership of the main lines. In imperial Germany, a nationalization (*Nationalisierung*) coincided with an integration of German land-state (*Länder*) railways into a German imperial railway. As the German Empire, *das Deutsche Reich*, was seen to be the German nation-state from the 1870s to World War I (WWI), a railway nationalization in Germany was equal to a "*Verreichung*," that is, a unification of railways under imperial control.

Secondly, Gotō and the body of text in the pamphlet itself about German railway organizations and their finance in this period paid little attention to the development toward railway unification after the failure of railway nationalization in Bismarck's age. He left out the coexistence of German land railways and their negotiations for (partial) unification till just before the WWI. In other words, Gotō did not compare Japanese success with German failure in railway nationalization simultaneously in a strict sense.

The confinement of the concept of German railway nationalization to only Bismarck's policy, seems to have led the railway historians in Japan to the confusion and identification of the German Empire with Prussia. Inspired by Friedrich Engels, some Japanese Marxist economic historians have seen Japanese railway nationalization through the lens of "Bismarckian nationalization" in which the needs of state power were emphasized more than economical rationality. The term *nationalization* was sure to be misleading when applied to the experience in the German Empire overall. This unconscious confusion of the German Empire with Prussia gave a distorted account of railway nationalization and unification in Germany. Although a large number of studies have been made on Prussian-German railway history over a long period of time, little is known about railway unification before the German National Railway (*Reichseisenbahn*) in 1920. This problem has been brought to light first by Yamada (Yamada 2001, 181–190). Because of Matsunaga (2008, 2012), Japanese studies on German railway nationalization at the turn of the century has caught up to the historical observations in Germany before WWII (Legatz 1920; Kittel 1941).

Much ink has been spent studying German railway nationalization' in the 19th century. There have been too many studies of traditional railway history that narrate the institutional development there. One of the sources of them is, for example, Kech's classical analysis (Kech 1911). Over the past few decades, there has been a renewal of interest in railway nationalization in Germany, and the historians have shed new light on the subject. Fremdling pointed out the vitality of private railway companies alongside the insignificance of the state-ownership of the development of German railways in the 19th century. "The state-ownership (*Verstaatlichung*) reproduced merely this inefficient traffic performance, only on another level." He also referred to nationalization and stated that, "(f)irst a state-ownership on the Reich-level, like the one planned by Bismarck, could have prevented this (inefficiency)" (Fremdling 1975, 60, fn 219). On the contrary, Kocka stressed the influence of German state bureaucracy in railway management (Kocka 1987). Following this research angle, Then, who examined the organization of German railways in detail, pointed out that the transition to state ownership was the last entrepreneur's decision (Then 1997). Ziegler followed the development of railway policy of German land-states and concluded that the birth of German state intervention in the process progressed from the predominance of private companies over state railways to the state-ownership of those railways (Ziegler 1996).

Despite of subtleties of those studies, the view that the failure of railway nationalization that has been established in the first half in the 20th century, is generally accepted by them. Even recent studies (Mette 1973; Ziegler 1996; Gall 1999; Mierzejewski 1999; Mitchell 2000; Yamada 2001) consider the failure of the nationalization plan of Bismarck primarily in the political context; that is, the particular dynamics imperial Germany. Certainly, in getting a good grasp of political and institutional development of German railway nationalization, the problem of German particularism cannot be overemphasized. It is commonly accepted that other German land-states and their governments claimed a degree of autonomy in railway policy and stuck to their position to the end. There seems to be much truth in this. As Mitchell has shown in full detail, "(i)n the German railway industry, particularism, not nationalism, was triumphant before 1890" (Mitchell 2000, 129).

But reflection on the development in the 19th century makes clear that central to the failure to make a national railway system was *not only* the problem of German particularism. Even when all German railways were owned by the *Reich*, (as Bismarck had wished at first, and which was rejected by the German land-state governments), the *Reich*-wide supervision of railways should have at least partially been realized, as the Imperial Constitution essentially determined this was to be so. The inability of the Imperial Railway Office *(Reichseisenbahnamt; REA)* to act as executive has to be investigated.

The foregoing argument also leaves unanswered the question of development of centralization in a land-state railway. The Prussian State railway, which was the representative for German land-state railways before the *Reichseisenbahn*, has been investigated by Mitchell, but their organizational reform after 1890 needs further consideration. The reform developed in the period of discussion between land-states concerning railway integration in the empire. One of the aims of organizational reform of the Prussian State railway seems to be a nationalization. Adding to this, particularism of various land-states seen from the Prussian side remains a matter for debate. To explore particularism in various state railways in detail will be helpful for our understanding of the German failure of railway nationalization in comparison with relative Japanese success.

Therefore, this study compares the railway nationalization in imperial Germany with the Japanese case.

This paper falls into three sections. In Sect. 6.2, we will try to describe the institutional development of German railway nationalization from 1870s to the beginning of the 20th century, the eve of the WWI. The development consists of two phases along a time series: the first one was the failure of the empire-scale integration, and the second was the progress of state ownership of railways in

Prussia and the institutional centralization in the Prussian State railway, which was seen at first as the follow-up of the failure of railway integration of the empire. We focus attention on the Railway Central Bureau (*Eisenbahnzentralamt; EZA*) that was established for centralized processing of materials and vehicles. In Sect. 6.3, the railway integration between the territory-states at the turn of the century are observed. We confirm the intersection of various intentions both inside and outside of the territory states, which appeared in the form of particularism (*Partikularismus*) that has been said to have impeded German railway integration. In Sect. 6.4, we compare Japanese railway nationalization with German cases. Our concern here is to clarify the similarities and differences between the German railway and that of Japan in the process of nationalization. It offers a key to understanding the global railway history from the angle of social relationships between the transportation sector and state.

6.2 Unifying Railways in Germany Before the First World War: From the Setback of Nationalization by Bismarck Until the Centralization of the Prussian Railway Administrative Organization

*6.2.1 Failure of the Imperial Railway Bureau (*Reichseisenbahnamt *[REA]) in the 1870s*

Otto Bismarck, who became the German Empire's minister in 1871, had the idea of a nationalized railway that was owned by the unified state, the German Empire (*Reich*). Seeking its own financial resources for the new German empire, Bismarck twice drafted a bill to transfer the local railway lines to the *Reich*. He established the Imperial Railway Bureau *(Reichseisenbahnamt*; REA) as the imperial office to be nationalization enforcement agency in 1873. Albelt von Maybach, a leading traffic administrative bureaucrat, took office as the secretary.

However, the Federal Council *(Bundesrat)* that gathered representatives from each regional state (*Land*) rejected the nationalization bill. From a financial point of view, the states could not let go of their railway. Also, the identity of a sovereign state with railways was to be adhered to within the federal system of the German empire. In response to their resistance, Bismarck abandoned the plan of railway unification under the government of the *Reich*. There is a common understanding that the *REA* after that remained in a small, almost powerless, organization while being a central government agency in the rail administration in the *Reich* (See, Kunz 1996).

As for the powerlessness of the government office, detailed observations on specific cases have not been made. It will be necessary to analyze the failure of nationalization again from this angle by the giving new light to the function of the REA (See, Banzawa 2012).

After his resignation as the President of the REA, Maybach had a very low evaluation of REA's performance, function, and its members' abilities. The influence of Maybach, which was maintained even after his resignation and the then his taking a position at the Prussian Ministry of Public Works, reached the internal personnel affairs of REA. In response to the promotion plan of engineer Emmerich in February 1883, and after expressing a negative view in a letter to Bismarck on the grounds of his personality and mediocre capability, he criticized strongly the current situation of REA itself as follows:

> If one does not fill the Wiebe's post temporary, as long as I consider, it will not be an official disadvantage. A right field is not found at present as a supposed far-reaching effectiveness of the REA, especially in the pure technical field. I must regard as a great important thing that the REA will keep in a closer contact with the Prussian Ministry that grasps all action and will try the best to avoid an unnecessary over production of documents (*überflussige Vielschreiberei*) that was the results of insufficient activity. (Maybach, A letter from Maybach to Bismarck (07.02.1883)).

The criticism of the REA's lack of cooperation with the Prussian State Railway accorded with Bismarck's reconsideration in 1877 after the discard of the second bill of the law for nationalization. Bismarck expressed his regret that it was the failure of the previous organization that the heading of the REA could have been identified with that of the Prussian State Railways (Kunz 1996, 15).

Based on this recognition, Maybach proceeded to acquire and nationalize the railways in Prussia as the head of the newly established Kingdom Ministry of Public Works. It seems reasonable to conclude that this nationalization of railways was the follow-up to the failure of railway integration in the *Reich* during the Bismarck era.

The nationalization process, which was the acquisition and state ownership of private railways in Prussia, formed the image of Bismarckian nationalization." In the study of railway history in Japan, this concept of Bismarckian nationalization has been understood as "railway nationalization with the main objective of bureaucratic control rather than economic necessity" based on the provisions of Friedrich Engels. It can be said that this concept has prevented the Japanese from understanding German railway history (Yamada 2001, 185–201).

What has been missing in such an understanding was the regionally decentralized organization in the established Prussian State Railway administration. According to the railway provision (1895), which was imperialized in 1894, the Prussian State Railway has established a decentralized organizational composition in which the regionalKingdom Railway Offices (*die Königliche Direktionen* [KED]) stood in parallel.

6.2.2 Establishment of the Prussian Railway Central Bureau (**Das Eisenbahnzentralamt**; *[EZA]*) *at the Turn of the Century*

The establishment of the Prussian Railway Central Bureau (*das Eisenbahnzentralamt*; EZA) in 1907, on the other hand, was a partial introduction of a centralized system of the management of Prussian State Railways.

At the turn of the century, with the expansion of the scale of Prussian State Railway (KED average route length from 1350 km in 1895 to 1700 km in 1907, and the number of employees from 14,385 to 21,359), each KED was forced to employ more technical experts than ever due to the increase of significance of mechanical technology. The necessity to reduce the burden on KEDs with more than 30,000 personnel was clear. But the division and reduction of the KED was considered undesirable for operational, expense, and defense reasons. As a result, a reform of the railway provision was inevitable (Breitenbach, A letter from Minister of Public Works (Breitenbach) to the King (20.11.1906)).

From the yearbook of Prussian government, we can show the history of the EZA (*Handbuch über den Preussischen Hof und Staat für das Jahr* 1908, 331–333). The EZA was established on March 28 in 1907 by the Ordinance of the Ministry of Public Works, based on "Railway Order on Supplemental Change of the State Railway Administrative Order and Installation of the Central Railway Bureau" on March 25. The EZA was established in the capital, Berlin, but it was treated in parallel with other KED (those in Berlin, Stettin, Posen, Breslau, Halle, Magdeburg, and Altona) as "Railway Office in Brandenburg" and was legally equal to KED. The EZA's missions were as follows;

- Adjustment and homogenization (*Ausgleich*) of railway cars (and passenger cars) at the National Railway Vehicle Alliance, arrangement preparation at each KED of new cars, substitution for vehicle ordering, settlement with foreign railway administration bureau on mutual use of vehicles, material adjustment;
- Procurement of important materials such as upper facility materials in general, locomotives, vehicles, trains, coal etc., inspection receipt of materials, supervision of construction, adjustment of used materials;
- Creation of a master plan regarding general facilities, making a uniform rule on general problems of sales conservation and sales activities in general;
- Mediation of intermediate officials (staff), as far as provisions concerning the entire region of the national railway are concerned;
- Duties related to the uniform payment of staff; and
- Workers' disease insurance and pension services.

The staffing composition of the EZA at the time of establishment, and their location and position before the appointment can be confirmed as follows.

The total number of EZA officials in 1907 was 86, of which 46 were employed by the Berlin head office, and four were employed by the attached facilities (the Railway Inspection Department). The first resident *(Präsident)* was Wilhelm Hoff

(1851–1940) who was the Director—General of Traffic Administration in the Ministry.

By comparison of the name registers in the yearbook of 1907 with those of 1908, the former job places and the number of the staff of the EZA headquarters staff are shown as follows. Berlin, 17 (including 2 in the headquarters); Rhein Province 8, Silesia Province 4; Frankfurt (a.M) 3; Westphalia 2; Saxony Province 1, Prussia 1 (*Handbuch über den Preussischen Hof und Staat für das Jahr* 1907, 202, 265. 294, 328–329, 387, 411, 463–464, 515, 517, 616, 653, 699–700, 761–763; 1908, 331–333). The regional balance of the eastern, western, and central part of kingdom was thought to be conscious by the authorities, but EZA officials were not selected from all of KEDs. Many transferred from the Berlin administration office and the Essen control office. Essen was the key point of the Ruhr heavy industrial area, and the office there was one of the gateways to a successive career course for high-ranking railway officials.

Out of 46 EZA officials, 37 were technicians. Of these, 34 inspectors who worked outside Berlin were technicians. There were 12 engineers whose previous workplaces were not the Berlin Control Authority. Likewise, four from Essen, two from Frankfurt am Main, and one each from Elberfeld, Saarbrücken, Bromberg, Katowitz, and Breslau worked at the EZA. Besides Berlin, there are many transfers from various administrations in the west.

From the above observations, it can be seen that concentrations of relatively high level technicians in Berlin had progressed to some extent from the beginning of EZA. In addition, those who were appointed from the heavy industrial zone in the west of the kingdom accounted for the majority of the technical officials above the inspector level in the EZA, supplementing this capital-centered staffing.

In short, EZA was added to the structure and organization of the decentralized railway administration of the Prussian Kingdom in order to concentrate the technical aspects of railway management work within the kingdom in the capital city, Berlin. An attempt was made to absorb and concentrate the technical knowledge that existed in the industrial areas of the western and central part of the kingdom. In the later appointment of Saare to the president of the EZA in 1911, his connection with heavy industrialists of the industrial city of Katowitz in Silesia was emphasized (Breitenbach, "Instruction from Minister of Public Works" (16.12.1911)). It should be noted that there was a recognition that technological capability had been amassed outside Berlin as the precondition for the intention to advance the centralization of technology.

The mission of the Prussian EZA was succeeded by the Railway Central Office of the *Reichsbahn*. In this respect, the existence of EZA was indeed in line with the integration and unification of railways in the German Republic since the 1920s.

It is difficult to determine how the EZA's activities contributed to integration and unification at the national level. One reason is that the EZA was established just before the outbreak of the First World War. Since the wartime regime (and resulting mobilization) should have had a great influence on the operation of the national railway, it cannot be confirmed whether the intended purpose of the EAZ in peacetime was achieved. Kittel (1941) appreciates the significant influence of WWI on the destruction of concept of a sovereign nation with its own railway.

To add to this, the scope of the EZA's mandate was naturally restricted to the Prussian-Hessian Railway Alliance. No cooperation with other regional (state) railways advanced before the WWI. For this reason, the contribution of the EZA to the integration of the German federal system can be discussed only in terms of its effect on the progress of integration on the territory of Prussian-Hessen Alliance. A measurement of its tickle-down effects obviously had a negative result. Each German state railway had a regional relationship with various heavy-industry players, and each relationship introduced various locomotives. It took ten years after the World War to eliminate this overly fractured system and its inefficiencies. The first locomotives which unified standards and parts at the national level, the Type 01 and 02, were developed from 1925 to 1926; that is, after the establishment of *Reichsbahn* in the Weimar Republic (Ebenfeld 2002, 38–39).

It is clear that the function of the EZA was also influenced by institutional negotiations between the countries.

6.3 Unifying Railways in Germany Before the First World War (2); Stagnation of Bilateral Negotiations and Particularism (*Partikularismus*)

As the Prussia State Railway occupied about two-thirds of the total length of the German Empire's railway line, its advantages were clear. Therefore, unification of the German railway industry should have been done mainly by Prussia.

Within the organization of each railway system (e.g., organizational structure, management materials) in the German Empire, the integration of standards and parts did not exist. There was no sharing of freight cars. If a train from the Prussian state railway from Berlin transported cargo to Munich, for example, the car would be emptied on the way back to Prussian territory. As a result, economic inefficiency was obvious. Though it was obvious that integration was needed for a wider geographic area, attempts at this did not go well.

6.3.1 Bilateral Negotiations Between German State Railways After the 1880s

According to Legatz (1920) and other official documents, one can organize the background of the bilateral negotiations as follows:

In 1897 the Prussian-Hessian Railway Community was established and was evaluated as successful from the viewpoints of its economic effect. Inspired by this success of community, the negotiations for management integration between rural areas began in the early twentieth century (Legatz 1920, 333–334).

In 1902, movement on alliance with Prussia by Württemberg Railway began. The aim was to search for a limited alliance, such as a freight car community, without losing railway sovereignty as a semi-independent state in the empire. In 1904, the discussion of management integration was resumed by the exchange of letters between the German emperor (Wilhelm II) and the king of Württemberg (Wilhelm II) who was his relative (Legatz 1920, 339–340; Wilhelm II, A Letter from the King of Württemberg to the Emperor (10.04.1904); Wilhelm II, A letter from German Emperor to the King of Württemberg (04.06.1904)). The management-material community was first conceived then. The Heidelberg conference included consultations with the four national railways, and the agreement was concluded (Legatz 1920, 342–343; Ausschrift aus der "Frankfurter Zeitung" (14.01.1904a)).

The Kingdom of Bavaria eventually denied, however, that their participation resulted in the community being "centralized." Bavaria was trying to bring its autonomy within the empire to the forefront and was known for continuing to insist strongly on the particularism (*Partikularismus*) in state railway policy. The claime that the Bavarian State Railway was not included in Article 42 of the Imperial Constitution that established the nationalization of the German railway was accepted to some extent by other land-states (Legatz 1920, 340). In response to the Freight Forwarding Centralization Agreement, Bavaria even showed its original plan (Breitenbach, A letter From Breitenbach to German Emperor (05. 05.1907)).

In 1909, the German State Railway Vehicle Association was formed, making vehicle use in freight transportation more efficient (Legatz 1920, 390). As progress in this area up to this point ended before World War I, the movement leading up to railway integration was brought up during the war. The fundamental solution to the problem was brought up after the establishment of the *Reichseisenbahn in 1920.*

6.3.2 Intention of Each Regional State and Actual State of Their Particularism

It has been emphasized that particularism (and specifically that of Bavaria) was an obstacle to German railway nationalization. But in order to accurately grasp the actuality of (in) decision-making within state railways, it should be confirmed by contemporary internal documents.

As a result of observing the document of the Prussian Commerce Ministry on the transition of the Prussian State railways to the German Empire, it can be seen that there were intersections of various intentions even in Prussia's government after 1904.

For railway integration on the German imperial scale, the Ministry of Public Works, which was the competent authority for transportation administration in the

Prussian Kingdom, was often reluctant. Even in consultation with Württemberg and Baden from 1904, the Prussian Minister of Public Works stated that Prussia's leadership in negotiations should never be destroyed and it would override integration progress (Breitenbach, A letter from Breitenbach to German Emperor (05. 05. 1907), 53–54).

There were also persistent voices that were concerned about the financial burden of railway integration. In 1906, Prussia decided to discontinue negotiations once "for financial reasons on the national law"(*Breitenbach*, A letter from Breitenbach to German Emperor (24.12.1911), 82). After the establishment of the Prussian-Hessian railway community, the profitability of railway management was certainly declining as far as the calculated operation ratio showed (Fremdling et al. 1995, 75, 207, 400–401).

It can be said that the mainstream bureaucrats of the Ministry of Public Works were reluctant to promote national railway integration. In his report to the emperor in December 1911, Breitenbach, the Prussian Minister of Public Works, even denounced a retired officer of the railway administration as the integration promoter. According to him, the negotiating representative, the former railway finance president, Kirchhof, who promoted the "railway community," was overly prominent at the Heidelberg conference in 1904 and the Berlin meeting in 1905. Britainbach claimed that Kirchhoff was not the representative of the ministry (Breitenbach, A letter from Breitenbach to German Emperor (24.12.1911), 83–84).

Contrary to the general image that railway nationalization was being promoted mainly by Prussia, there was also an independent attitude that should be called "particularism" in the Prussian State Railway.

Of course, there were variations in opinion among the insiders of each country, and the intentions of various groups conflicted in the negotiations between the countries. From the Kingdom of Prussia, movement of various interest groups in other land-states was observed as follows:

Though only the Mine-Neckar railway from the Baden State Railway had joined the Prussian-Hessian Railway Community in 1902, opinions were complicated over this experience. The conservatives in Baden vigorously warned that their railway lines would be swallowed by the Prussian State Railway (Eisendecher, A letter from the Prussian Legation for Baden to Minister of Foreign Affairs (Schillingsfürt) (03.05.1899).) Meanwhile, they agreed to promote the participation in the railway community for fiscal reasons and did not neglect to warn against the Württemberg Railways. Stealing a march on them the Prussian state tried to merge the Württemberg Railways with the Prussian State Railway (Eisenbahn- und Tarifpolitisches. *Badisch Landpost* no.79 (06.04.1899)).

While the government in Württemberg was concerned that the influence of the Prussian National Railway would increase, economic circles such as the Chamber of Commerce and others aggressively argued for railway integration for financial reasons. They announced they were anxious that the appropriate timing will be missed by the government's sticking to railway sovereignty (Wöllwarth 1899; Süddeutschland und die Eisenbahnen. *Die Post* no.219 (11.05.1911)).

The Saxon State Railway fought the so-called "Railway War" with Prussian Railways and continued to compete on a parallel routes. But some Saxons considered the difference in size with the Prussian State Railway and made an argument that management integration was advantageous from a financial perspective (Eisenbahnabschlusse. *Dresdner Journal* (27.01.1904b).). It was also claimed one benefit would be that the main trunk line would pass through Saxony if the railway lines were connected (Die deutsche Eisenbahngemeinschaft.Leipziger *Tagesblatt* (04.01.1902).). In Saxony, attention was paid to the opinion from Stuttgart that claimed that Prussia would not actively encourage the other states to integrate railways and therefore the other land state should move spontaneously. The particularism in South Germany was criticized in this article (Deutsches Reich: Breslau, 14. Mai Zur Reichseisenhan-Frage. *Schlesische Zeitung* no.337 (15. 05. 1901)).

Even in Bavaria, which was regarded as the most radical representative of particularism, there were voices claiming the economic significance of railway integration, and in part, compared the situation to the German Customs Union in the 19th century (Zur Frage der Vereinheitlichung des deutschen Eisenbahnwesens *Münchener-Neuesten Nachrichten*, no.194 (26.04.1901)).

There certainly existed a particularism, and it seemed to defeat the progress of nationalization. But, it must be said that its manifestation was extremely multifaceted.

In addition, opinions inside German states changed over time. After the outbreak of the First World War, the German chiefs of staff called for railway integration across the *Reich*. Acting in favor of this sentiment, the Prussian Ministry of Public Works took the initiative. In 1915, the Ministry of Public Works proposed the transfer of Prussian State Railways to the empire, whereas the differences in opinion among ministries and agencies within the Prussian government were revealed. Among the government officials who agreed to promote a railway transfer were the Minister of Public Works, Minister of Agriculture and Forestry, Minister of Education, Minister of Education, Minister of Justice, Minister of Commerce and Industry, economists, and the Minister of State, K. Helfferich, who worked in the postwar movement of railway integration, the establishment of the *Reichseisenbahn*. On the other hand, the Minister of Finance, the Minister of Home Affairs, and the Minister of Military opposed the transfer of the railway to the empire. While they shared a sense of the possible benefits of increasing management and transportation efficiency through railway integration, financial consideration for losing their own railway were also great. Speculation on the self-interest of various ministries and agencies in the kingdom have been found (For example, Helfferich, Äusserung des Staatsministers Dr. Helfferich über die Gründung eines Deutschen Reichseisenbahnnetzes).

Legatz (1920) has pointed out the evils of particularism with respect to German railway nationalization, but we should pay attention to the fact that this classic overview also views the "particularism" in each state railway in multiple ways. Legatz mentioned the variation of organizational composition, various standards, facilities, and profitability in individual state railways at the end of the nineteenth century as problematic. Among them, inconsistency in "personnel affairs (*Personalwesen*)" has been pointed out. Between German state railways in the empire with different traditions, there was no unification of uniforms and ranking chapters, and the awareness about coworkers was rare. Also, since organizational structure also varies according to the size of the business, the distinction between the senior (upper), intermediate (middle) and lower classifications of the staff and the work content were diverse in each railway; even the salary systems were not standardized. Indeed, from the mid-19th century, the Association of German Railway Administrations (*Verein für Deutschen Eisenbahn Verwaltungen*; VDEV) had unified various standards as the railway standards of Germany, and in principle, the operational organizations required for a unified railway industry could not differ so much from each other. But a fine difference in organizational human resources management naturally occurred. And the small differences were decisively important to the parties. For example, the staff of an intermediate class of the Bavarian State Railway had much fewer duties, which was impossible for their colleagues in Prussian State Railway (Legatz 1920, 334). Such a difference was a serious problem between those in the field. The personnel at the operational site were not motivated at all for national integration. The integration of individual railways was, to some extent, accompanied by large costs for systematic adjustments and that was well-known by railway authorities in each country.

6.4 Failure of Nationalization in the German Empire—A Comparison with the Japanese Case

The failure to unify and integrate railways in the German Empire can be summarized as follows:

There was the absence of a system that would facilitate railway integration and an executing agency centered in *Reich*. On the other hand, the sentiments of the local states toward the railway, which have been said to prevent nationalization of the German railways, were multifaceted and also fluid, including those espoused by the Prussian Kingdom. The "particularism" has been thought to be the consideration of the benefits in securing a railway as a financial source coupled with the concept of a sovereign state with independent control of a railway. At the same time, it can be pointed out that there was also a consciousness about the cost of organizational adjustments that would necessarily have been included in railway integration. In short, there was an institutional difficulty in breaking the constraints imposed by the federal system in the German empire.

By comparing the failure of German railway nationalization with the success of railway nationalization after the publication of "Railway Nationalization Act" (*Tetsudō-kokuyū-ho*) in 1906 (Meiji 39) in Japan, let us consider similarities and differences between these two cases in order to shed a new light on their experiences.

In Japan, the Temporary Road State Initial Preparation Bureau (*Rinji-tetsudō - kokuyū-junbi-kyoku*) was set up as an executive body for the acquisition of private railways and nationalization based on the Railway Nationalization Act in 1906. It completed 17 private railway acquisitions in a shorter period than planned.

> As expected, there were various requests regarding the determination of the purchase price from each company, . .but a solution was found under cooperative negotiations(Tetsudō-Shō (ed) 1920, 13–14).

In 1908 (Meiji 41), along with the expansion of the government railway in Japan, the Agency of Railways directly under the cabinet was established by integrating the Railway Bureau (*Tetsudō-kyoku*) in the Ministry of Communications (*Teishin-Shō*) and the Agency of Imperial Railways (*Teikoku-Tetsudō-Chō*). As a result of this establishment, the operation, which had conventionally been done "in accordance with centralism" in Japan, was refined, and the whole country was divided into five railway provinces: *Hokkaido* (*Hokkaidō* and *Seikan* sea-route jurisdiction in Sapporo), an eastern part (*Tōhoku* region jurisdiction in *Ueno*), *Chūbu* (*Tōkaidō* line and central line jurisdiction in *Shimbashi*), a western part (*Sanyō* line · *Shikoku* and *Kampu* sea route in *Kōbe*), and *Kyūshū* (*Kyūshū* Jurisdiction in *Moji*). Each railway management office supervised each regions transportation offices, maintenance lines offices, and factories.

It can be said, on the other hand, that the integration of railway management at the imperial-level was not realized in the end. While the Agency of Railways also took part in the supervision of the management of the South Manchuria Railway, the railways in colonial Taiwan were under the jurisdiction of Transportation Bureau of Taiwan Sōtokufu. This happened despite the strong influence of Shinpei Gotō, who became the first president of the Agency of Railways. In 1906, when the Railway Nationalization Act of Japan was promulgated, all the Korean railways that the Japanese had built were also publicized. But the management of then state-owned railways in Korea were put under the Railway Bureau of *Chōsen-Sōtokufu*. It is also evident that the subsequent railway management in the Japanese colonies and Manchuria were largely unaware of the integration with the Japanese national railway on the mainland (see Kō 1999; Lim 2001, 2013; Tsai 2016).

Which organization in the German railway system should we compare with the Agency of Railways in Japan in terms of the jurisdiction and the function?

It is obvious that the REA does not correspond to it. As we have seen above, since Bismarck's railway integration had not been realized in the 1870s, REA stayed within imperial government agencies with a very small "*Reichsbahn*" on imperial territory in Alsace-Lorraine. In addition to this, the objective of railway integration that the German imperial government originally assumed was the integration of railways of each local state. But the Japanese imperial government did not merge railways owned by other sovereign authorities. The Japanese experience of railway nationalization was not comparable to the plan in the German empire in terms of scale. It was confined to only the acquisition of the railway in the Qing Dynasty during the early stage of Taiwanese colonial rule (in 1895) and the partial transfer of the southern Manchuria branch of the Chinese Eastern Railway from Russia (in 1905). As mentioned above, a plan of integration of colonial railways did not have any relation with the railway nationalization initiative in Japan.

Needless to say, it is difficult to find examples in Japan that were similar to the negotiations between local states and the intersection of various intentions inside and outside the respective localities in Germany at the turn of the century that we have observed above. If we were to pursue the possibility of comparing such experiences between Germany and Japan, we could not avoid observing the overseas colonies of the Japanese empire. The unification of the operation of colonial railways was, however, a challenge for Japanese governmental authorities in the wartime regime during the Second World War. But the fact the unification of railways throughout the Japanese empire did not come true as a result of the defeat of World War II cannot be overemphasized. Considering the failure of railway integration in the German empire, one of the imperial nation-states in Europe where the local states shared linguistic- and cultural identity and traditional federal ties, it seems quite probable that a railway integration in the Japanese empire would fail in some way, even if the Japanese empire survived after the Second World War.

The railway nationalization in Japan corresponded, to some extent, to the progress in state (or governmental) ownership of private railways in Prussia under the command of Maybach after the failure of railway nationalization (in the meaning of imperial state ownership, "*Verreichung*") by the REA. The newly established Prussian Ministry of Public Works proceeded with the acquisition of private railways, and in the 1890s after nearly achieving state-owned operation, Karl von Thielen, Minister of the Public Works, reorganized the management organization throughout the kingdom: twenty railway administration bureaus had jurisdiction over each region. One can safely say that the function of Prussian Ministry of Public Works corresponded to that of the Japanese Agency of Railways. From this point, we can reevaluate here that the claim of "Bismarckian nationalization" of Japanese railways in the research of Japanese economic history was accurate as an object of comparison (Oikawa 2005).

Naturally, we should pay attention to the difference between the two bureaus. Subtle differences in organizational structure (cannot be ignored. Each Prussian KED was under the command of the President (*Präsident*) and were under the direct control of the Minister of Public Works. The organization corresponding to the Agency of Railways directly under the cabinet in Japan was the Second Bureau within the ministry that was functionally responsible for the railway administration. The relatively newly established EZA was a centralized organization, while it was institutionally placed in a way that did not destroy the framework of traditional decentralized administration in Prussia. The EZA was legally equivalent to the KEDs and was stated as one of the "railway bureaus in Brandenburg" in the bureaucratic order that can be seen from an official publication (*Handbuch über den Preussischen Hof und Staat für das Jahr 1908, 331–333*). Its function had great weight in terms of technology, and the internal department of Japanese Agency of Railways (such as the Construction Department) overlaps the EZA's functionally.

It seems, generally, that the attempt to promote integration in terms of technology was to deal with future problems such as large scale operation and high-speed mass transportation. In this respect, the installation of the Agency of Railways by the Japanese railway nationalization process and installation of the EZA in Prussia responded to the same problem. From that viewpoint, the internal department of the Agency of Railways—the Engineering Department (the Construction Office and the Manufacturing Office)—and the EZA are counterparts. One may say that both of them are contemporary organizations in global railway history.

Regarding the control of railway technicians, the Japanese Imperial (National) Railway was identical to the Prussian State Railway in terms of the need for coordination between the central and the regional units. In Japan, during the expansion of governmental railways and private railway acquisition, the engineers in local railway companies established their own technical system in each district. For this reason, it was a challenge to cope with the difference between the technical systems of the public and the private railways after nationalization. At the same time, it was necessary to change generations of railway engineers (Nakamura 1998, 204–208; 2015, 209–210). The same problem of technical unification applied to German-Prussian and imperial railways.

The Agency of Railways (*Tetsudō-In*) was very different from the EZA in that the relations between the industrial centers divided locally and the local railway administrations (the KEDs) of each region was not considered as prerequisites for technical development that was centralized and prompted by a central bureau. Due to the organizational revision in 1913 (Taishō 2), the Agency of Railways directly controlled the factories which had been under the jurisdiction of each local railway bureau (Tetsudō-shō 1920, 16–18). While these factories were returned to the affiliated KEDs in a short period of time, personnel rotation between various local railway bureaus and the Agency of Railways (later Construction Bureau in the Ministry of Railways) had the significance of on-the-job training and bureaucratic career paths for individual technical officials (Tetsudō-gijyutsu-kenkyūsho Gojyusshūnen-shi Kankō-iinkai 1957, 1–14).

Certainly, as we have seen above, the Japanese National Railway Administration had also the problem of governance of railway technicians and engineers with their respective backgrounds in each area.

But there was a phenomenon that the technical knowledge itself flowed in one way from the capital to the provinces in Japan, and we should emphasize that aspect in comparison with Germany. One cannot say that the training of Japanese railway engineers was closely related to industrial technology education in each region. On the other hand, Prussia had the nuclei of industrial centers in western and central Germany with regional leading technical education institutions (various Technical Colleges (*Technische Hochschulen*)) in areas other than the capital, Berlin. What this distinction makes clear (again) was that the difference between Japan and Germany in positioning on the economic development stage or their difference in character as an "industrialized country". That influenced the development of railway administrative organization. The importance of connecting with several regional heavy-industrial centers found in Prussian or the German railways was one of the reasons for inconsistency of capital equipment and delay in standardization. What has to be looked more carefully is the influence of the difference in material procurement and in the initial source of import of material and technology in each formerly private railways in Japan.

The process of preparing railway statistics at the imperial level by REA indicate its lack of organizational structure that was necessary for collecting regional information as a central government agency and the difficulties in collaboration among ministries and agencies (Maybach, A letter from Minister of Public Works (Maybach) to REA (12.09.1879); See, Banzawa 2012). The Agency of Railways was free of the former problem (a lack of hierarchical and regionally dense organization) by having the railway district and local bureaus.

On the other hand, the railway administration in Japan divided before the railway nationalization and establishment of the Agency of Railways. The point to observe here is whether the division of railway policy between the Railway Department under the Ministry of Communications and the Agency of Imperial Railways under the Ministry of the Interior caused the inefficiency of governmental railway administration.

Adding to this, the cost of organizational adjustment was important when comparing the two railway nationalizations. From that point of view, the passive attitude toward inter-state integration at the level of the staff in the railway organization in Germany must be examined in comparison with the Japanese case. It is necessary to reconfirm how the formerly private Japanese railway staff responded during the railway nationalization. The substantial penetration of socialism and labor movements in the railway industry in Germany and the rights movement of the workers, even if they disguised a non-political form, should be kept in mind when comparing the two railway nationalizations (Banzawa 2006, 298–316).

6.5 Preliminary Conclusive Remarks

We may go on from this to the preliminary conclusion as follows:

There has been a substantial difference in the meaning of "railway nationalization" between Japanese and German railway economic histories. It is remarkably difficult to regard the unification of the governmental railways operated by multiple (semi-)sovereign states that had been already established (such as in the German case in the same light as the achievement of the state acquiring of private railways (as in Japan). It is probably the process of expansion of the Prussian State Railway that can be discussed as integration in the sense of the state-owning railway nationalization in Japan. The question then arises about the tradition of the federal system of decentralization of the Prussian kingdom, the inadequate centralization corresponding to it, and the absence of an organization like the Agency of Railways.

All of this leads to the conclusion that the existence (or absence) of the federal system was a decisive factor in the way railway nationalization took place.

Regarding the influence that the federal system and its traditions had on the development of the railway industry, it is necessary to make some degree of relativization from two directions.

First, it must be noted that the functions of a centralized and unified institution depended on the execution of organizational management. The cases of the REA and the EZA illustrate the significance of actual operation of bureaus.

Secondly, one may say that the operational organization of the official railway sector of Japan followed that of Prussia in a reverse path and reached an equilibrium point. In Prussia, the EZA, an organization that intended to centralize railway administration to some extent, was added to the decentralized organizational structure. In Japan, meanwhile, a reorganization was repeated after the railway nationalization in which a certain degree of decentralized organizational structure was adopted corresponding to the technical limit of centralized operation (Sakiyama 2013, 80). At the turn of the century, the period when the superior position that the railway industry maintained was beginning to weaken, the necessity of large-scale high-speed mass transport has to be reconsidered in comparison with other new traffic infrastructure. The increase in efficiency of institutional functions in railway sector was pursued with priority, apart from each national railway's historically determined frameworks. It is quite likely that Gotō's remark that we introduced at the top of this article, was an expression not of his feeling of satisfaction, but of his intention to challenge to the simultaneous problems in the railway sector.

Primary Sources

Breitenbach, Paul Justin. 20. November 1891. A letter from Minister of Public Works (Breitenbach) to the King (20.11.1906) (*Geheimes Staatsarciv Preussischer Kulturbesitz, Berlin-Dahlem* (hereafter *GStA*), HA Rep.89, No.29483 Eisenbahnzentralamt 1906–17).

Breitenbach, Paul Justin. 5. May 1907. A letter From Breitenbach to German Emperor (05. 05.1907). (*GStA* I. HA Rep. 89, no.29472/1 Übergang der Preussischen Staatseisenbahnen auf das Deutsche Reich und den gemeinsamen bundesstaatlichen Eisenhanbetrieb 1876–1911, 53–62).

Breitenbach, Paul Justin. 16. December 1911. Instruction from Minister of Public Works (16.12.1911). (*GStA* I. HA Rep.89, no.29483 Eisenbahnzentralamt 1906–17).

Breitenbach, Paul Justin. 24. December 1911. A letter from Breitenbach to German Emperor (Wilhelm II.) (24.12.1911).(*GStA* I. HA Rep. 89, no. 29472/1 Übergang der Preussischen Staatseisenbahnen auf das Deutsche Reich und den gemeinsamen bundesstaatlichen Eisenhanbetrieb 1876–1911, 81–91).

Eisendecher, (Karl). 3. May 1899. A letter from the Prussian Legation for Baden (Eisendecher) to Minister of Foreign Affairs (Schillingsfürt) (03.05.1899). (*GStA* III HA Ministerium der auswärtigen Angelegeheiten II no.7092 Akten betreffend die Eisenbahnbeziehungen zwischenden deutschen Bundesstaaten, soweit sie nicht in Spezialakten behandelt werden).

Helfferich, Karl. 7. February 1916. Äusserung des Staatsministers Dr. Helfferich über die Gründung eines Deutschen Reichseisenbahnnetzes. (*GStA* III HA Ministerium der auswärtigen Angelegeheiten II no. 7103 Akten betreffend die Gründung eines deutschen Reichseisenbahnnetz (1915-1916)).

Maybach, Albert. 12. September 1879. Übersicht der fürdie Aufstellung der statistischen Nachrichten von den Preussisichen Eisenbahnen an die Bahn-Verwaltungen ergangenen Bestimmungen (A letter from Minister of Public Works(Maybach) to Reichseisenbahnamt (12.09,1879)). (*Bundesarchiv* (*hereafter BA*), R/4101 Reichseisenbahnamt 1 Abteilung A (Allgemeines) 477 Aufstellung einer Eisenbahnstatistik für das Deutsche Reich Bd.2).

Maybach, Albert. 7. February 1883. A letter from Maybach to Bismarck (07.02.1883) (BA, R43 Reichskanzlei (»Alte Reichskanzlei«),1447 Akten betreffend die Mitglieder und höhren Beamten des Reichseisenbahnamtes, Bd.12, 10 (1–3).)

Wilhelm II (King of Württemberg). 10. April 1904. A Letter from the King of Württemberg (Wilhelm II.) to the Emperor (Wilhelm II.)(10.04.1904). (*GStA* I.HA Rep. 89, no. 29472/1 Übergang der Preussischen Staatseisenbahnen auf das Deutsche Reich und den gemeinsamen bundesstaatlichen Eisenhanbetrieb 1876–1911, 44–46).

Wilhelm II (German Emperor). 4. June 1904. A letter from German Emperor (Wilhelm II) to the King of Württemberg(Wilhelm II) (04. 06.1904). (*GStA* I. HA Rep. 89, no. 29472/1 Übergang der Preussischen Staatseisenbahnen auf das Deutsche Reich und den gemeinsamen bundesstaatlichen Eisenhanbetrieb 1876–1911, 39–41).

References

Banzawa, Ayumu. 2006. *Doitsu-kōgyōka ni okeru tetsudōgyō (Railways in German Industrialization)*. Tokyo: Yūhikaku.

Banzawa, Ayumu. 2012. What caused the failure of nationalisation of the railway system in Germany?: malfunction of the German Imperial Railway Office (Reichseisenbahnamt) in the 1870s and 1880s. Osaka University, Graduate School of Economics and Osaka School of International Public Policy (OSIPP) 12–18, Discussion Papers in Economics and Business, http://www2.econ.osaka-u.ac.jp/library/global/dp/1218.pdf. Accessed 27 Dec 2016.

Ebenfeld, Stefan. 2002. Einheitsloks und neue Ordnung. Im Dienst von Demokratie und Diktatur: Die Reichsbahn 1920–1945, ed., DB Museum, Nürnberg: DB Museum.

Fremdling, Rainer. 1975. Eisenbahnen und deutsches Wirtschaftswachstum, 1840–1879. Ein Beitrag zur Entwicklungstheorie und zur Thorie der Infrastructur. Gesellschaft für Westfälische Wirtschaftsgeschichte. Dortmund: Gesellschaft für Westfälische Wirtschaftsgeschichte.

Fremdling, Rainer, Ruth Federspiel, and Andreas Kunz (eds.). 1995. *Statistik der Eisenbahnen in Deutschland 1835–1989*. St.Katharinen: Scripta Mercaturae Verlag.

Gall, Lothar. 1999. Eisenbahn in Deutschland: Von den Anfängen bis zum Ersten Weltkrieg. In *Die Eisenbahn in Deutschland: Von den Anfängen bis zur Gegenwart Gebundene Ausgabe*, ed. Lothar Gall, and Manfred Pohleds. München: Carl Beck.

Handbuch über den Preussischen Hof und Staat. 1907. Berlin.

Handbuch über den Preussischen Hof und Staat. 1908. Berlin.

Kech, Edwin. 1911. *Geshichte der deutschen Eisenbahnpolitik*. Leipzig: Göschen'sche Verlagshandlung.

Kittel, Theodor. 1941. Der Plan eines, Bundes der deutschen Staatseisenbahnen', das Ende des preussisch-sächsisich Eisenbahnkriegs (Ein Beitrag zur Vorgeschichte der Reichseisenbahnen). *Archiv für Eisenbahwesen* 64: 537–550.

Kō, Sonbon. 1999. *Shokuminchi tetsudō to minshū seikatsu: Chōsen, Taiwan, Chū goku-tōhoku (Railways and Popular Life in Japanese Colonies: Korea, Taiwan and Northeast China).* Tokyo: Hōsei-daigaku Shuppankai.

Kocka, Jürgen. 1987. Eisenbahnverwaltung in der industriellen Revolution:Deutsch Amerikanische Vergleiche. In Historia Sociallis et Oeconomica. Festschrift für W. Zorn zum 65. Geburtstag, VSWG Beiheft 84, eds., Hans Kellenbenz and Hermann Pohl, 259–277 Wiesbaden/Stuttgart: Franz Steiner.

Kunz, Wolgang. 1996. Vom Reichseisenbahnmat (1830–1919)zur Gegenwart. *Jahrbuch für Eisenbahngeshichte.* 28: 5–26.

Legatz, L. 1920. Zur Geschichte des Reichseisenbahngedankens. *Archiv für Eisenbahwesen* 43 (321–350): 617–635.

Lim, Chaisung. 2001. Senjiki shokuminchi Chōsen ni okeru rikuuntōsei no tenkai: kokutetsu-yusōno keikakuka wo chūshin to site (The South Manchuria Railway Company's Development of the Railway Business: An Examination of Efficiency and Profitability). *Tochiseido Shigaku (The Journal of Political Economy and Economic History)* 43–2: 1–17.

Lim, Chaisung. 2013. Mantetsu ni okeru tetsudōgyō no tenkai: kōritsusei to shūekisei no shiten yori (The evolution of land transportation control in colonial Korea during the war 1937–1945). *Rekishi to Keizai* (The Journal of Political Economy and Economic History) 55 (4): 1–15.

Matsunaga, Kazuo. 2008. Doitsu ni okeru tetsudō no "kokuyūka" toha: futatsu no "kokuyūka" wo megutte (What was railway nationalization in Germany?: about two "nationalization"). *Tetsudō Shigaku (The Journal of Railway History)* 25: 70–73.

Matsunaga, Kazuo. 2012. Doitsu ni okeru tetsudō no chūōshūkenka to ryōhō: tōitsu ni mukatte senro ha tsuduku (The Centralization of railways and Land-states in Germany; Rails leads to the integration). In *Tetsudō* (*Kindai yōroppa no tankyu 14*) (Railways (Explorations into Modern Europe 14)), eds. Takeshi Yuzawa, Shigeru Koike, Toshihiro Tanaka, Kazuo Matsunaga and Kiyoyuki Ono. Kyoto: Minerva Shobō.

Mette, Thomas. 1973. Der Übergang der deutschen Ländereisenbahnen auf das Reich, Ph.d. Dissertation. Humboldt-Universität Berlin.

Mierzejewski, Alfred. 1999. *The Most Valuable Asset of the Reich.* Chapel Hill and London: The University of North Carolina Press.

Minami Manshū Tetsudō Tōa Keizai Chōsa Kyoku (The East Asia economic research bureau in South Manchuria Railway Company) (1909), *Doitsu ni okeru testudō no tokubetsusoshiki oyobi sono zaisei: Tōa Keizai Chōsa Kyoku Zassan Kō-1gō* (The Special organization and its financial policy in Germany: miscellaneous collection of the East Asia economic research bureau in South Manchuria Railway Company Ko-1). Tokyo: Minami Manshū Tetsudō Tōa Keizai Chōsa Kyoku.

Mitchell, Allan. 2000. *The Great Train Race: Railways and the Franco-German Rivalry, 1815–1914.* New York: Berghahn.

Nakamura, Naofumi. 1998. *Nihon tetsudōgyō no keisei 1869–1894 nen (The formation of Japanese Railway industry).* Tokyo: Nihon Keizai-Hyōron-Sha.

Nakamura, Naofumi. 2015. Kindai-Nihon no tetsudō-gijyutsusya (Railway engineer in modern Japan). In *Genba-syugi no kokusaihikaku* (International comparison of site-preferential principles), ed. Akifumi Taniguchi, 181–212. Kyoto: Minerva-shobō.

N.N. 1899. Eisenbahn- und Tarifpolitisches. *Badisch Landpost* no.79 (06.04.1899) (*GStA* III HA Ministerium der auswärtigen Angelegeheiten II no.7092 Akten betreffend die Eisenbahnbeziehungen zwischenden deutschen Bundesstaaten, soweit sie nicht in Spezialakten behandelt werden).

N.N. 1901. Deutsches Reich: Breslau, 14. Mai Zur Reichseisenhan-Frage. *Schlesische Zeitung* no.337 (15. 05.1901) (*GStA* III HA Ministerium der auswärtigen Angelegeheiten II no.7093 Akten betreffend die Eisenbahnbeziehungen zwischenden deutschen Bundesstaaten, soweit sie nicht in Spezialakten behandelt werden).

N.N. 1902. Die deutsche Eisenbahngemeinschaft. *Leipziger Tagesblatt* (04.01.1902). (*GStA* III HA Ministerium der auswärtigen Angelegeheiten II no.7093 Akten betreffend die Eisenbahnbeziehungen zwischenden deutschen Bundesstaaten, soweit sie nicht in Spezialakten behandelt werden.).

N.N. 1904. Ausschrift aus der "Frankfurter Zeitung" no.14. Erstes Morgenblatt, 49 (14.01.1904). (*GStA* I. HA Rep. 89, no.29472/1 Übergang der Preussischen Staatseisenbahnen auf das Deutsche Reich und den gemeinsamen bundesstaatlichen Eisenhanbetrieb 1876–1911).

N.N. 1904. Eisenbahnabschlusse. *Dresdner Journal* (27.01.1904). (*GStA* III HA Ministerium der auswärtigen Angelegeheiten II no.7094 Akten betreffend die Eisenbahnbeziehungen zwischenden deutschen Bundesstaaten, soweit sie nicht in Spezialakten behandelt werden).

N.N. 1911. Süddeutschland und die Eisenbahnen. *Die Post* no.219 (11.05.1911). (*GStA* III HA Ministerium der auswärtigen Angelegeheiten II no.7093 Akten betreffend die Eisenbahnbeziehungen zwischenden deutschen Bundesstaaten, soweit sie nicht in Spezialakten behandelt werden).

Oikawa, Yoshinobu. 2005. Tetsduo-kokuyūka no rekishiteki katei ni tsuite: Shima Yasuhiko "Nihon shihonsyugi to kokuyū-tetsudō" to Nakanishi Kenichi" Nihon shiyū tetudō shi" (On the historical process of Railway Nationalization in Meiji Japan). *Rikkyo-Keizaigaku-Kenkyū (Rikkyo Economic Review)* 58–4: 305–314.

Sakiyama, Tetsuo. 2013. Taisyō-syoki ni okeru kokutetsu (Tetsudō-In) no soshiki to gijyutsu yosei ni tuite: Taishō 5 nen kokutetsu "kansha-tōkyū oyobi otsu-gō ika kansha hyōjyun" no seiritsu haikei (About the organization and engineer training of Japanese National Railway (Japanese Imperial Railway Bureau) in the early Taishō-period). *Nihon-kenchiku-gakkai Tōhoku-shibu Kenkyu-hōkoku-shū (Report of Researches in the Tōhoku Branch of Architectural Society Japan)* 76: 79–82.

Tetsudō Gijyutsu Kenkyūsho Gojyushūnenshi Kankō Iinkai. 1957. *Gojyūnenshi (50th anniversary of Railway Technology Research Institute)*. Tokyo: Kenyūsha.

Tetsudō-shō (Japanese Imperial Ministry of Railway) (ed). 1920. *Kokuyū jūnen: honpō tetsudō kokuyūgo no shisetu narabini seiseki* (Railway nationalization in Japan), Tokyo: Tetsudō-shō.

Then, Volker. 1997. Eisenbahnen und Eisenbahnunternehmer in der Industriellen Revolution. Ein preussisch/deutsch - englischer Vergleich. Göttingen: Vandenhoeck & Ruprecht.

Tsai, Lung-Pao. 2016. Nihon tōchiki ni okeru Taiwan sōtokufu tetsudō no nanshinseisaku; Shin-koku Kantōshō Chōsen tetsudō no jirei (The general governor of Taiwan's southward advance policy: a case study of Chao Chow-Swatow Railway in Guangdang province of Qing Dynasty). *Rikkyo-Keizaigaku-Kenkyū (Rikkyo Economic Review)* 69–5: 1–24.

Wöllwarth, Freiherr Georg von. 1899. Preussisch-Hessische Eisenbahngemeinschaft. *Schwäbischer Merkur* no.149 (30.03.1899); *Staatsanzeiger für Württemberg*, no.79 (07.04.1899) (*GStA* III HA Ministerium der auswärtigen Angelegeheiten II no.7092).

Yamada, Tetsuo. 2001. *Doitsu shihonsyugi to tetsudō (German capitalism and railways)*. Tokyo: Nihon-Keizai Hyoronsha.

Ziegler, Dieter. 1996. *Eisenbahnen und Staat im Zeitalter der Industrialisierung. Die Eisenbahnpolitik der deutschen Staaten im Vergleich.* Stüttgart: F. Steiner.

Index

M. Sawai (ed.), *The Development of Railway Technology in East Asia in Comparative Perspective*, Studies in Economic History,
DOI 10.1007/978-981-10-4904-0

MIX
Papier aus verantwortungsvollen Quellen
Paper from responsible sources
FSC® C105338

If you have any concerns about our products,
you can contact us on
ProductSafety@springernature.com

In case Publisher is established outside the EU,
the EU authorized representative is:
Springer Nature Customer Service Center GmbH
Europaplatz 3, 69115 Heidelberg, Germany

Printed by Libri Plureos GmbH
in Hamburg, Germany